A Comprehensive Summary of the Benford's Law Phenomenon

On the Unequal Spread of Digits within Scientific and Typical Data

A Comprehensive Summary of the Benford's Law Phenomenon

On the Unequal Spread of Digits within Scientific and Typical Data

Alex Ely Kossovsky

World Scientific

NEW JERSEY · LONDON · SINGAPORE · GENEVA · BEIJING · SHANGHAI · TAIPEI · CHENNAI

Published by

World Scientific Publishing Co. Pte. Ltd.

5 Toh Tuck Link, Singapore 596224

USA office: 27 Warren Street, Suite 401-402, Hackensack, NJ 07601

UK office: 57 Shelton Street, Covent Garden, London WC2H 9HE

Library of Congress Cataloging-in-Publication Data
Names: Kossovsky, Alex Ely, author.
Title: A comprehensive summary of the Benford's law phenomenon : on the unequal spread of
    digits within scientific and typical data / Alex Ely Kossovsky.
Description: Singapore ; Hackensack, NJ ; London : World Scientific, [2025] |
    Includes bibliographical references and index.
Identifiers: LCCN 2024053277 | ISBN 9789819801282 (hardcover) |
    ISBN 9789819801299 (ebook for institutions) | ISBN 9789819801305 (ebook for individuals)
Subjects: LCSH: Benford's law (Mathematics) | Place value (Mathematics)
Classification: LCC QA273.6 .K67 2025 | DDC 519.2/4--dc23/eng/20241211
LC record available at https://lccn.loc.gov/2024053277

British Library Cataloguing-in-Publication Data
A catalogue record for this book is available from the British Library.

For any available supplementary material, please visit
https://www.worldscientific.com/worldscibooks/10.1142/14054#t=suppl

Desk Editors: Aanand Jayaraman/WSE-Gabriel Rawlinson

Typeset by Stallion Press
Email: enquiries@stallionpress.com

About the Author

Alex Ely Kossovsky is an independent scholar and the author of the books ***Benford's Law:*** *Theory, the General Law of Relative Quantities, and Forensic Fraud Detection Applications*, World Scientific Publishing Company, 2014; ***Small is Beautiful:*** *Why the Small is Numerous but the Big is Rare in the World*, Kindle Direct Publishing, 2017; ***Studies in Benford's Law:*** *Arithmetical Tugs of War, Quantitative Partition Models, Prime Numbers, Exponential Growth Series, and Data Forensics*, Kindle Direct Publishing, 2019; and ***The Birth of Science***, regarding Kepler's Celestial Data Analysis, Galileo's Terrestrial Experiments, and Newton's Grand Synthesis, Springer Publishing, 2020. Kossovsky is the inventor of a patented mathematical algorithm for data fraud detection analysis, registered with the US Patent Office. The author specialized in applied mathematics and statistics at the City University of New York and in physics and pure mathematics at the State University of New York at Stony Brook.

Contents

About the Author   v

Introduction   xiii

Section I: The Digits Phenomenon   1

Chapter 1:   The First Digit on the Left Side of Numbers   3

Chapter 2:   Benford's Law and the Predominance
of Low Digits   7

Chapter 3:   Second-Digit and Third-Digit Distributions   13

Chapter 4:   The Quantitative Origin of the
Digital-Numerical Phenomenon   17

Chapter 5:   The Scale Invariance Principle   23

Chapter 6:   The Base Invariance Principle   27

Chapter 7:   Physical Order of Magnitude of Data   29

Chapter 8:   Robust Measure of Physical Order of Magnitude   31

Chapter 9:   Two Essential Requirements for Benford Behavior   33

Chapter 10: Sum of Squared Deviations Measure   37

Chapter 11: The Mistaken Use of the Chi-Square Test
in Benford's Law   41

Section II: Causes and Explanations   45

Chapter 12: Multiplication Processes Lead to Positive
Skewness and Often to Benford   47

Chapter 13: Addition Processes Lead to the
Symmetrical Normal and away from Benford   53

Chapter 14: The Multiplicative Central Limit Theorem
and Lognormal Distribution   57

Chapter 15: Multiplications are More Prevalent than
Additions in Real-Life Data   61

Chapter 16: Tugs of War between Addition and Multiplication   63

Chapter 17: Partitions Typically Lead to Positive
Skewness and Often to Benford   69

Chapter 18: One-Dimensional Random Staged Partition   73

Chapter 19: One-Dimensional Chaotic Repeated Partition   77

Chapter 20: One-Dimensional Random Real Partition   81

Chapter 21: Two-Dimensional Random Partition   85

Chapter 22: The General Requirements for Partitions
to Converge to Benford   89

Chapter 23: Benford Model for Planet and Star Formations   91

Chapter 24: Consolidation and Fragmentation Processes   95

Chapter 25: Random Exponential Growth Leads
to Positive Skewness and Benford   99

Chapter 26: Data Aggregation Leads to Positive
Skewness and Often to Benford 103

Chapter 27: Chains of Statistical Distributions and
Benford's Law 109

Chapter 28: Meta-Explanation or the Explanation of all
Explanations 117

Section III: The Logarithmic Perspective 119

Chapter 29: Benford's Law as Uniformity of Mantissa 121

Chapter 30: Rising or Falling Mantissa Distributions 129

Chapter 31: Uniqueness of k/x Distribution and Its
Central Role in Benford's Law 131

Chapter 32: Related Log Conjecture 137

Chapter 33: The Random and Deterministic Flavors
in Benford's Law 145

Chapter 34: The Great Prevalence of the Digital
Development Pattern in Data 151

Chapter 35: The Absence of the Digital Development
Pattern in k/x Distribution 155

Chapter 36: Benford's Law in Its Purest Form 157

Chapter 37: Constant Base Raised to a Random Power 163

Section IV: General Results 167

Chapter 38: General Results in Benford's Law 169

Chapter 39: First Two Digits versus Last Two Digits 181

Section V: The Law of Relative Quantities   **185**

Chapter 40: The Related Concepts of Digits, Numbers,
             and Quantities   187

Chapter 41: The Arbitrariness of Our Positional
             Number System   189

Chapter 42: Two Radically Different Interpretations of
             the Benford Phenomenon   191

Chapter 43: The Quest for a Universal and
             Number-System-Invariant Measure   195

Chapter 44: The Shape and Nature of Histograms are
             Number-System Invariant   199

Chapter 45: Constructing a Three-Bin Histogram
             Signifying Small, Medium, and Big   203

Chapter 46: Constructing a Set of Infinitely Expanding
             Histograms   209

Chapter 47: Numerical Consistency in Bin Schemes for
             15 Real-Life Data Sets   213

Chapter 48: The Postulate on Relative Quantities   217

Chapter 49: Application of the Postulate via Generic
             Bin Scheme on k/x 221

Chapter 50: The Infinite Sequence Result for the Bin
             Scheme on k/x 227

Chapter 51: The General Law of Relative Quantities   229

Chapter 52: Benford's Law as a Special Case and Direct
             Consequence of GLORQ   233

Chapter 53: The Universal Law of Relative Quantities   237

Chapter 54: Benford Second-Order Digits Interpreted
             as an Irregular Bin Scheme   239

Chapter 55: Concluding Historical and Conceptual
        Perspectives   243

Appendix A: Infinite Sequence Reduction   247

Appendix B: Data Sets   251

Appendix C: Glossary of Frequently Used Abbreviations   253

Appendix D: Bibliography   255

Index   259

Introduction

The Benford's Law Phenomenon

Numbers represent physical quantities in the real world, as well as abstract values and counts of human interest and concern, and are recorded via our digital language system by expediently utilizing the ten digits from 0 to 9, just as text and books written in English, French, or German conveniently utilize the 26 letters from A to Z to constitute all written material. While the letters Z, X, and Q are quite rare in English text, the letters E, T, and O are very common and occur relatively frequently. Clearly, letters are not evenly and uniformly spread in text.

Surprisingly, and against all common sense, defying our innate intuition, the spread of the 10 digits within the numbers of real-life, scientific, physical, and typical data is also not uniform and equal but rather highly skewed and uneven. Benford's Law states that the first digits on the leftmost side of numbers are decisively unequal and that occurrences of low digits, such as 1, 2, and 3, are much more frequent than occurrences of high digits, such as 7, 8, and 9, as digital proportions decrease dramatically from around 30.1% for digit 1, being the most frequent and popular first digit, to around 4.6% for digit 9, being the rarest first digit.

The explanation of the phenomenon in extreme generality is that low digits represent small quantities while high digits represent big quantities; and since the small almost always outnumbers the big in the world, this is so for digits as well, so that low digits outnumber high digits. In other words, the vast majority of real-life scientific and typical data sets are positively skewed, where the small is numerous and the big is rare, so that histograms are mostly falling to the right in the aggregate from the minimum to the maximum, and consequently, overall, low digits have an advantage over high digits.

Empirical examinations of real-life data overwhelmingly confirm the existence of such uneven quantitative proportions in favor of the small. There are more poor people with small bank accounts than rich people with big bank accounts. There are more small planets and stars than big ones in the Cosmos. In geological data, there are more small rivers than big rivers in the world, and there are more harmless small earthquakes than devastating big ones. There are more villages than towns, more towns than cities, and more cities than metropolises. There are by far many more small creatures than big creatures in the biological world. There are only about 2 million big whales swimming in the oceans. There are 7 billion humans living on Earth, who are intermediate in size. There are over 300 billion small birds flying in the sky. Tiny little ants are even more abundant, with an estimate of over 100 trillion of them walking the Earth.

The vast list of topics and data types obeying this quantitative law of nature in favor of the small confirms the fact that the phenomenon is nearly universal. Mathematicians naturally attempt to numerically quantify this quantitative phenomenon in order to obtain an exact measure indicating by how much the relatively small is more numerous than the relatively big, and the generic approach is to focus purely on the primary quantities instead of the secondary digits.

Surprisingly, a certain quantitative numerical measure is found, with the property that its value is consistent and nearly universal across almost all data sets with sufficient variability having a high order of magnitude. This nearly universal pattern across almost all data sets is termed "the general law of relative quantities", or simply GLORQ as its acronym. As it happens, Benford's Law regarding the consistent digit distribution in almost all data sets with sufficient variability is demonstrated to be simply a mere consequence and a special case of GLORQ.

Benford's Law is found to be valid in almost all real-life statistics, such as in data relating to physics, chemistry, astronomy, geology, biology, economics, finance, accounting, engineering, and governmental census information. As such, Benford's Law constitutes a unique common thread running through and uniting data sets relating to all scientific disciplines, connecting and unifying the sciences with this common digital-numerical feature.

Awareness in the general public and academic research regarding this phenomenon have been steadily increasing in recent years and decades, with an explosion in the number of published articles regarding its manifestations in all fields of the sciences, as well as attempts at alternative mathematical and even conceptual explanations of the phenomenon.

Applications

In addition, interest in the applicability of Benford's Law as an excellent tool in data fraud detection is gaining recognition. Data cheaters and fraudsters are typically not aware of this digital pattern, and they often tend to concoct numbers quite randomly when inventing fake data; consequently, their digits are typically uniformly and equally spread in the approximate. The mathematician could easily detect the fraud by observing that the digits are not skewed in favor of lower ones as they are supposed to be. In particular, the phenomenon has enabled analysts in recent years to examine election data in order to ascertain whether vote counting was honestly reported or fraudulent in nature. Data forensics applications via Benford's Law in accounting, auditing, economics, and general financial reporting have been put into practice for over 30 years, beginning around the mid-1990s, and this is increasingly becoming nearly a routine check on data authenticity.

The History of the Discovery

The phenomenon was first discovered by Simon Newcomb (1835–1909), who was a Canadian-American astronomer and mathematician known for his talent in dealing with immense amounts of numerical data. In 1881, Newcomb published an extremely short two-page, one-sheet article titled "Note on the Frequency of Use of the Different Digits in Natural Numbers", in which he correctly described and tabulated the nine proportions of the first digits,

Simon Newcomb

known today as Benford's Law, but without explicitly writing an analytical algebraic expression for this. Newcomb nonetheless mathematically stated the law indirectly, in complete generality, via the concept of the logarithm of numbers. His article was subsequently forgotten and ignored for six decades until Benford aroused interest in the phenomenon with his acclaimed article just a year prior to the outbreak of the Second World War.

Frank Benford

Frank Benford (1883–1948) was an American physicist and electrical engineer who published 109 articles in the fields of optics and mathematics and was granted 20 patents on optical devices. Benford, who did not know of Newcomb's earlier work, independently rediscovered the phenomenon and, in 1938, wrote a lengthy article about it titled "The Law of Anomalous Numbers". As opposed to Newcomb, who for the most part just stated the fact with a broad mathematical statement without offering forensic evidence, Benford explicitly stated the algebraic expression for the probability of the first digits and set out to empirically examine 20 different large collections of data types, systematically recording their digital results and compatibility with the law. Subsequently, the phenomenon gained attention, albeit ever so slowly and gradually over the decades since Benford's publication.

Both discoverers, Newcomb and Benford, were led to the discovery by the same physical observation, which indirectly hinted at the patterns in occurrences of digits within numbers of actual data. What caught their attention was the highly differentiated physical wear and tear of pages in older books of tables of logarithm values, which were in that era commonly used by scientists and engineers for arithmetic calculations, well before the advent of the computer and the calculator. The visual condition of these books was such that it showed the first pages pertaining to the first digits 1, 2, and 3 to be more strained and worn out by use, but increasingly less so throughout the book for higher digits, with the last pages pertaining to the first digits 7, 8, and 9 seeming to be in relatively good condition, as if these last pages weren't much in use.

It was natural for Newcomb and Benford to infer that the scientists and engineers looking up the tables of these logarithmic books were more often in need of using them for numbers relating to low first digits in the first pages and more rarely so for numbers relating to high first digits in the last pages, reflecting the actual unequal spread of first digits in real-life data.

Newcomb's short article starts by referring to this differentiation in the usage of the pages:

> *That the ten digits do not occur with equal frequency must be evident to any one making much use of logarithmic tables, and noticing how much faster the first pages wear out than the last ones.*

Benford's long article also starts by referring to this differentiation in the usage of the pages:

> *It has been observed that the pages of a much used table of common logarithms show evidences of a selective use of the natural numbers. The pages containing the logarithms of the low numbers 1 and 2 are apt to be more strained and frayed by use than those of the higher numbers 8 and 9. Of course, no*

> *one could be expected to be greatly interested in the condition*
> *of a table of logarithms, but the matter may be considered more*
> *worthy of study when we recall that the table is used in the*
> *building up of our scientific, engineering, and general factual*
> *literature.*

About This Book

This book has been written for two distinct audiences. The book
and the topic are certainly intended for expert mathematicians,
statisticians, and scientists, as well as university students of the
respective disciplines, who could naturally and immediately relate
to and grasp all the chapters and parts of the book and who will
find the book to be thought-provoking and highly informative. But
the book has also been written for the layman, lifelong learners, the
non-expert, and the educated general public, who are not necessarily
proficient in mathematics, statistics, or the sciences, but who will
still be able to understand the topic overall. The author's emphasis
on visual graphs and drawings, as well as the focus on conceptual
narratives, will enable the general non-professional reader to skip the
more challenging mathematical parts without any loss of continuity
and without diminishing their ability to comprehend the essence of
the Benford phenomenon.

This book represents a sincere and concentrated attempt by the
author to narrate this digital, numerical, and quantitative story
of the Benford's Law phenomenon as simply and as concisely as
humanly possible while still ensuring a comprehensive coverage of
all its aspects, results, causes, explanations, and perspectives. The
author has dedicated the past two decades to research on Benford's
Law, published three books in this field, and invented a mathematical
algorithm for related data fraud detection via an inner and more
refined digital pattern, which earned an official and valid patent from
the US Patent Office.

The aim of this fourth book by the author is to arrive at a clear
synthesis of all known aspects of the phenomenon, narrating a more
decisive and integrated view as opposed to merely writing a brief
collection of all the previous three books put together, and in the
spirit of the saying attributed to the philosopher Aristotle perhaps,
that the whole is greater than the sum of its parts.

Since this book is a summary of the Benford's Law phenomenon, aiming at a concise exposition requiring significantly less time and effort from the readers, the author's focus is on the main results and explanations while excluding the supporting mathematical proofs and the more detailed explanations, all of which are fully explored and outlined in the previous three books. An additional emphasis of this book is on the connection of the law to concrete real-life data sets and, in particular, to the physical and scientific manifestations of the law, instead of focusing solely on pure mathematical formulas and abstractions to the exclusion of all else. For researchers interested in an in-depth study of Benford's Law and GLORQ, the previous three books would prove immensely useful in obtaining a better understanding.

Suggestions, questions, feedback, as well as constructive criticism about the content of the book would be sincerely welcomed by the author and could be sent to the email address akossovsky@gmail.com.

SECTION I
The Digits Phenomenon

Chapter 1

The First Digit on the
Left Side of Numbers

It has been discovered that the first digit on the left side of typical numbers in the world is frequently of low values, such as $\{1, 2, 3\}$, but relatively rarely of high values, such as $\{7, 8, 9\}$. Let us demonstrate this numerical tendency in real-life data by examining the molar mass values of 32 randomly chosen chemical compounds, namely molecules, as shown in Figure 1. The molar mass is defined as the weight (in grams) of a huge collection (mole) of a given molecule. One mole is 6.022×10^{23}, hence the molar mass is the weight of 6.022×10^{23} such identical molecules.

Molecule	Formula	Molar Mass
Caffeine	$C_8H_{10}N_4O_2$	194.191
Glucose	$C_6H_{12}O_6$	180.156
Water	H_2O	18.015
Hydrogen Peroxide	H_2O_2	34.015
Citric Acid	$C_6H_8O_7$	192.124
Morphine	$C_{17}H_{19}NO_3$	285.343
Lactose	$C_{12}H_{22}O_{11}$	342.296
Potassium Cyanide	KCN	65.116
Sodium bicarbonate	$NaHCO_3$	84.007
Ethanol	C_2H_5OH	46.068
Ozone	O_3	47.998
Pentane	C_5H_{12}	72.149
Lithium Metaborate	$LiBO_2$	49.751
Zinc Nitride	Zn_3N_2	224.154
Iron Sulfide (Pyrite)	FeS_2	119.975
Nitrogen Tribromide	NBr_3	253.719
Silver Cyanide	AgCN	133.886
Sodium Peroxide	Na_2O_2	77.978
Sodium Chloride (salt)	NaCl	58.443
Silver Oxide	Ag_2O	231.736
Barium Nitrate	$Ba(NO_3)_2$	261.337
Sodium Azide	NaN_3	65.010
Titanium Nitride	TiN	61.874
Sodium Fluoride	NaF	41.988
Aluminium Oxide	Al_2O_3	101.961
Lead Nitrate	$Pb(NO_3)_2$	331.210
Silver Dichromate	$Ag_2Cr_2O_7$	431.724
Sodium Carbonate	Na_2CO_3	105.988
Iodine Monobromide	IBr	206.808
Sodium Chlorite	$NaClO_2$	90.442
Cadmium Niobate	$Cd_2Nb_2O_7$	522.631
Zinc Tungstate	$ZnWO_4$	313.247

Figure 1. Chemical Formula and Molar Mass of 32 Molecules

194.191	84.007	133.886	101.961
180.156	46.068	77.978	331.210
18.015	47.998	58.443	431.724
34.015	72.149	231.736	105.988
192.124	49.751	261.337	206.808
285.343	224.154	65.010	90.442
342.296	119.975	61.874	522.631
65.116	253.719	41.988	313.247

Figure 2. Molar Mass of 32 Random Molecules

194.191	**8**4.007	**1**33.886	**1**01.961
180.156	**4**6.068	**7**7.978	**3**31.210
18.015	**4**7.998	**5**8.443	**4**31.724
34.015	**7**2.149	**2**31.736	**1**05.988
192.124	**4**9.751	**2**61.337	**2**06.808
285.343	**2**24.154	**6**5.010	**9**0.442
342.296	**1**19.975	**6**1.874	**5**22.631
65.116	**2**53.719	**4**1.988	**3**13.247

Figure 3. First Digits of the Molar Mass of 32 Random Molecules

Figure 2 depicts this data set of 32 molar mass numbers, whereas Figure 3 emphasizes with bold font and black color the first digits on the leftmost side of these 32 numbers. The ordered digits are:

$$\{1, 1, 1, 3, 1, 2, 3, 6, 8, 4, 4, 7, 4, 2, 1, 2, 1, 7, 5, 2, 2, 6, 6, 4, 1, 3, 4, 1, 2, 9, 5, 3\},$$
$$\{1, 1, 1, 1, 1, 1, 1, 1, 2, 2, 2, 2, 2, 2, 3, 3, 3, 3, 4, 4, 4, 4, 4, 5, 5, 6, 6, 6, 7, 7, 8, 9\}.$$

The first digits in this molar mass sample consist mostly of low values. Indeed, 18 numbers start with the low digits $\{1, 2, 3\}$, while only four numbers start with the high digits $\{7, 8, 9\}$.

A summary of the first-digit configuration of the molecule sample is given as follows:

Digit index: $\{1, 2, 3, 4, 5, 6, 7, 8, 9\}$

Digit count totaling 32 values: $\{8, 6, 4, 5, 2, 3, 2, 1, 1\}$

Proportions of digits with "%" sign omitted: $\{25, 19, 13, 16, 6, 9, 6, 3, 3\}$

The website (www.convertunits.com/compounds/) provides the chemical formula and the molar mass of some 2175 known compounds, focusing on those used in heavy industry, the pharmaceutical industry, the food industry, and the metallurgical industry, as well as many naturally occurring and synthetic compounds. The above sample of 32 molecules was chosen from this set of 2175 chemicals in an almost random manner, except perhaps for some slight subliminal personal bias toward organic and well-known molecules, but without mentally focusing on the first digits in any conscious way.

Examination of the entire set of 2175 compounds and their first digits yields the highly skewed and uneven proportion vector of $\{31.9, 25.2, 16.1, 8.4, 5.7, 4.3, 2.9, 3.2, 2.3\}$, with the "%" sign omitted. The result of this relatively big data size strengthens the perception that this phenomenon is real and that the observed uneven digit frequencies of the above 32 molecules is intrinsic and not due to some very rare selection of molecules, where by chance the first digits just happen to be skewed in favor of low digits.

Could the result of this empirical data analysis related to chemistry be extrapolated to other types of data in the world as well as to other scientific disciplines? Does it indicate a definite general pattern of digits within numbers of real-life data? Yes, indeed this is the case! This digital pattern is found in the vast majority of real-life data sets. Hence, one may then conclude with the phrase "**not all digits are created equal**", or rather "not all first digits are created equal", even though this fact of life seems to be contrary to intuition and against all common sense.

Chapter 2

Benford's Law and the Predominance of Low Digits

Benford's Law states that

Probability[First Digit is d] = $LOG_{10}(1 + 1/d)$

$LOG_{10}(1 + 1/1) = LOG(2.00) = 0.301$
$LOG_{10}(1 + 1/2) = LOG(1.50) = 0.176$
$LOG_{10}(1 + 1/3) = LOG(1.33) = 0.125$
$LOG_{10}(1 + 1/4) = LOG(1.25) = 0.097$
$LOG_{10}(1 + 1/5) = LOG(1.20) = 0.079$
$LOG_{10}(1 + 1/6) = LOG(1.17) = 0.067$
$LOG_{10}(1 + 1/7) = LOG(1.14) = 0.058$
$LOG_{10}(1 + 1/8) = LOG(1.13) = 0.051$
$LOG_{10}(1 + 1/9) = LOG(1.11) = 0.046$

- - - - - - -

1.000

Figure 4 depicts the distribution, while Figure 5 visually depicts Benford's Law as a bar chart.

Digit	Probability
1	30.1%
2	17.6%
3	12.5%
4	9.7%
5	7.9%
6	6.7%
7	5.8%
8	5.1%
9	4.6%

Figure 4.   Benford's Law: First Digits

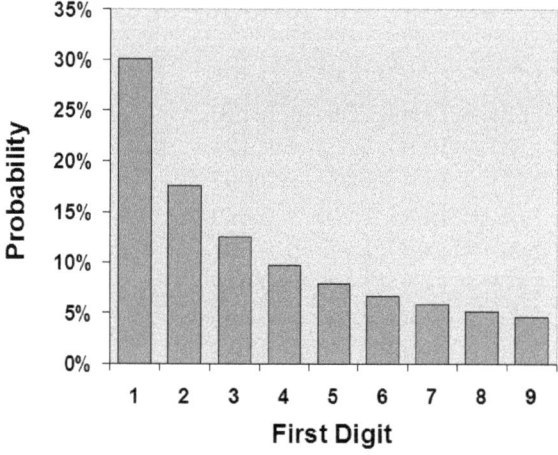

Figure 5.   Benford's Law: Probability of First Digit Occurrences

Remarkably, Benford's Law is confirmed and found in the vast majority of real-life data sets. The law applies to data sets with high variability (having a high order of magnitude) and which are positively skewed (the histogram is asymmetrical, falling on the right side). These two data features are nearly universal and are almost always found in data relating to physics, chemistry, astronomy, economics, finance, accounting, geology, biology, engineering, governmental census data, and nearly all other types of data. Benford's Law therefore constitutes the only common thread running through and uniting data relating to all scientific disciplines, connecting and unifying the sciences with this common digital-numerical feature.

Data sets are never exactly as in $\text{Log}(1 + 1/d)$, since there is always some small or even noticeable deviation from it, but the first digits are fairly close to the theoretical Benford configuration, as seen in the case of 2175 chemical compounds. The law only specifies the general trend and pattern, in probabilistic terms, although the vast majority of physical data sets occurring in nature of very large sizes (say tens or hundreds of thousands of values) and with a very large order of magnitude (say over 4 or 5) are extremely close to the ideal Benford configuration, with only tiny deviations from it. It should be noted that these nine logarithmic values of Benford's Law in the expression $\text{LOG}_{10}(1 + 1/d)$ are irrational numbers, as most logarithmic numbers are, hence in a formal mathematical sense, none of them could equal exactly a finite ratio of integral values expressing the proportion of the first digit d in any given data set, such as [data points where the first digit is d]/[total number of data points in the data set], which is a rational number.

Data sets provided by individuals, companies, and organizations having digital configurations that significantly deviate from the $\text{Log}(1 + 1/d)$ proportions come under the strong suspicion of being fake, fraudulently concocted, and intentionally invented, unless they are of very small size, say less than 50 data points, or unless they are of low variability, namely with a low order of magnitude of less than 2 in the approximate.

Since the vast majority of real-life data sets are such that they are positively skewed and are of sufficient variability, having a high order of magnitude, the Benford phenomenon is ubiquitous and nearly always relevant. For example, the population values of cities and towns of any given country typically vary greatly, from perhaps 25 residents for small villages to mega cities with over 1,000,000 residents. Here, the high order of magnitude is calculated as $\text{LOG}_{10}(1{,}000{,}000/25) = 4.6$. In addition, population data are highly skewed in favor of the small. There are more villages than towns, more towns than cities, and more cities than metropolises. Therefore, population data sets are Benford. Relatively rare exceptions are heights, weights, and IQs of people, which are not at all Benford because they suffer from very low variability as well as from symmetry (non-skewness), being of the Normal distribution. For example, the heights of all students at a very large university vary from, say, 1.35 meters for the shortest student to the tallest, proud and accomplished basketball player of 2.15 meters. Hence, the order of magnitude is calculated as $LOG_{10}(2.15/1.35) = 0.2$, which is very low. In addition, the height data appear as symmetrical non-skewed Normal. Surely, for this student height data, the first digit is nearly always 1 and only rarely 2, while digits 3 to 9 never occur in the first place, therefore student height data is clearly non-Benford.

For fractional numbers such as, say, 0.00913, the focus here is actually on the first meaningful digit, or "**First Significant Digit**", counting from the left side of numbers, excluding any possible encounters of zero digits which merely signify ignored exponents in the relevant set of powers of 10 in our number system. Hence, for 0.00913, the first significant digit is 9, and the proceeding zeros are ignored, as they are considered "insignificant". For the number 2365, the first digit is 2, and 2 represents the greatest quantity of 2,000, while 3 represents the lower quantity of just 300, and so forth. Clearly, 2 bears the greatest quantitative weight for the number 2365, hence implicitly the term "significant" here also refers to the most significant quantitative digit, which is of course the first digit. The first digit 2 in the number 2365 is also said to "lead" the other digits of {3, 6, 5}, being placed ahead of them, hence the alternate name of "**First Leading Digit**". For the lone integer 8, the leading digit is simply 8. For negative numbers, the negative sign is ignored, thus, for −715.9, the leading digit is 7.

For data where all values are non-fractional and equal to or greater than 1, namely $X \geq 1$, such as in the above example of the molar mass data, the first digit literally on the leftmost side of any number is always non-zero, and it is also the first leading digit and the first significant digit, and necessarily one of the nine digits {1, 2, 3, 4, 5, 6, 7, 8, 9}.

Note: $LOG(X)$ or $Log(X)$ notation refers to our decimal base 10 number system by default, hence the more detailed notation of $LOG_{10}(X)$ is often — but not always — avoided. The natural logarithm base e is referred to as $LOG_e(X)$ or simply as $\ln(x)$.

A useful expression yielding the first leading digit of any positive number X > 0 is given by

First digit of $X = \mathrm{INT}(X/10^{\mathrm{INT}(\mathrm{LOG}(X))})$

The INT function refers to the integer just below X or to X itself if X is exactly an integer, which is formally called the "floor of X", being the greatest integer less than or equal to X.

For example, INT(7.2) is 7, INT(9) is 9, INT(0.85) is 0, and INT(−5.2) is −6. The LOG function in this formula refers to the decimal logarithm of X with base 10. This expression for the first digit should prove to be quite useful in computer implementations of Benford's Law.

This expression is easily generalized to any positional number system base B by simply substituting B for 10, and by referring to base B LOG function instead of base 10 decimal LOG function. The generalized expression is then:

First digit of $X = \mathrm{INT}(X/B^{\mathrm{INT}(\mathrm{LOG}_B(X))})$

Chapter 3

Second-Digit and Third-Digit Distributions

Figure 6 emphasizes in bold font and black color the second digits from the leftmost side of the numbers of the molar mass of 32 random molecules.

1**9**4.191	8**4**.007	13**3**.886	10**1**.961
1**8**0.156	4**6**.068	7**7**.978	33**1**.210
1**8**.015	4**7**.998	5**8**.443	43**1**.724
3**4**.015	7**2**.149	23**1**.736	10**5**.988
1**9**2.124	4**9**.751	2**6**1.337	20**6**.808
2**8**5.343	2**2**4.154	6**5**.010	9**0**.442
3**4**2.296	1**1**9.975	6**1**.874	5**2**2.631
6**5**.116	2**5**3.719	4**1**.988	3**1**3.247

Figure 6. Second Digits of the Molar Mass of 32 Random Molecules

Evidently, the second digits of the molar mass data are more equally distributed, having only a mild tendency to favor low digits over high digits. Indeed, 15 numbers are with the low second digits of {0, 1, 2, 3}, while 11 numbers are with the high second digits of {6, 7, 8, 9}. There is no dramatic contrast between the second digits here, merely a slight bias toward low digits.

A summary of the second-digit configuration of the molecule sample is given as follows:

Digit index: $\{0, 1, 2, 3, 4, 5, 6, 7, 8, 9\}$

Digits count totaling 32 values: $\{4, 4, 3, 4, 3, 3, 2, 2, 4, 3\}$

Proportions of digits with "%" sign omitted: $\{13, 13, 9, 13, 9, 9, 6, 6, 13, 9\}$

For the second, third, and all higher digital orders, digit 0 is also included in the distributions, respected and considered as a proper digit in spite of its emptiness, but digit 0 is not part of the first order. The second leading digit (second from the left) of 7834 is digit 8, that of 0.003591 is digit 5, and that of 4093 is digit 0. The third leading digit (third from the left) of 3271 is digit 7. The fourth leading digit (fourth from the left) of 0.0981054 is digit 0. As we shift our focus from the very left corner of any multi-digit number, moving our attention gradually toward the right, examining the digits composing the number, and as soon as we encounter our first significant non-zero digit, we then start including all digits, even empty zero ones. This is why for the number 0.0981054, the first two zeros are meaningless for us, and they are totally ignored, but then we encounter digit 9, and this digit is respected and declared as being the first digit and quite significant, then 8 is declared as the second, 1 is declared as the third, and then as we encounter the 0 digit, it is respected and declared as the fourth digit. Readers might consider these two seemingly contradictory rules regarding the zero digit as arbitrary and perplexing, but these are not rules or decrees, but rather this simply arises from the cyclical ways the digital orders occur as we move to the right along the x-axis from the zero origin toward positive infinity, as shall be demonstrated more clearly in later chapters. Readers are welcome to anticipate all these by contemplating the relationship between standing on any particular X point along the positive x-axis and the digital configuration (regarding all orders) of that X number. It's a challenging task and better left for later chapters.

In addition to its decisive statement about the first digital order, Benford's Law also gives consideration to higher digital orders, stating their probabilistic distributions. According to Benford's Law, there exist some slight probabilistic dependencies between the orders: in essence, a slight positive correlation between them. In extreme generality, the *conditional* second-digit order has more skewness and is less equal (in favor of low digits) whenever the first digit is low, and it is a bit less skewed and more equal (in favor of high digits) whenever the first digit is high.

For example, for a near-perfect Benford data set of very large size, assembling only the numbers beginning with digit 1 (i.e., conditional on the first digit being 1) would yield second digits as follows:

Benford's Law second assuming first is 1:

{13.8, 12.6, 11.5, 10.7, 10.0, 9.3, 8.7, 8.2, 7.8, 7.4}.

But, assembling numbers beginning with digit 9 (i.e., conditional on the first digit being 9) would yield second digits as follows:

Benford's Law second assuming first is 9:

{10.5, 10.4, 10.3, 10.2, 10.0, 9.9, 9.8, 9.7, 9.6, 9.5}.

Unconditional second-digit probabilities refer to the proportions of the second digits in relation to the entire data set, containing all possible first digits in their natural Benfordian proportions.

The *unconditional* probabilities according to Benford's Law for the second, third, and fourth orders are as follows:

BL 2nd: {11.97, 11.39, 10.88, 10.43, 10.03, 9.67, 9.34, 9.04, 8.76, 8.50}

BL 3rd: {10.18, 10.14, 10.10, 10.06, 10.02, 9.98, 9.94, 9.90, 9.86, 9.83}

BL 4th: {10.02, 10.01, 10.01, 10.01, 10.00, 10.00, 9.99, 9.99, 9.99, 9.98}

Digit distribution for the second-digit order is only slightly skewed in favor of low digits compared to the much more dramatic skewness of the first-digit order, where digit 1 is about six times more likely to occur than digit 9. Digit distribution for the third-digit order shows only tiny deviations from the 10% possible equality. Figure 7 depicts the superimposed charts of first-, second-, and third-digit distributions according to Benford's Law. As even higher orders are considered, such as the fourth or the fifth orders, digits rapidly approach digital equality, attaining for all practical purposes the uniform and balanced 10% proportions for all the 10 possible digits of 0−9.

Second-Digit and Third-Digit Distributions

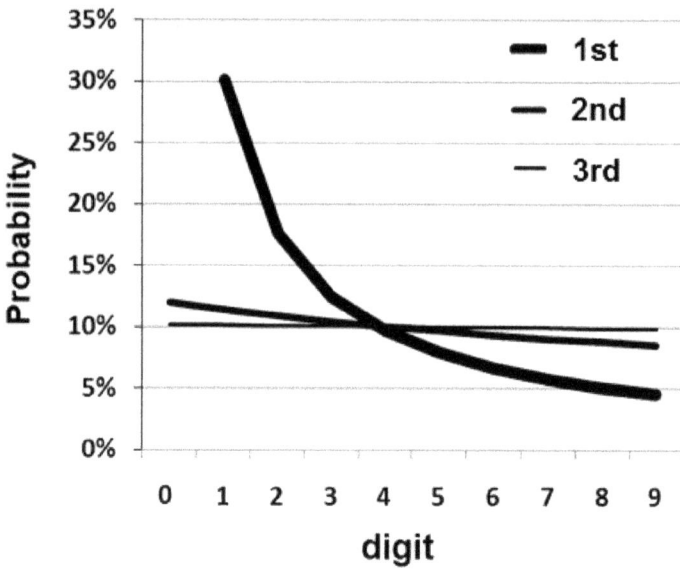

Figure 7. Benford's Law: Charts of 1st, 2nd, and 3rd Digits Superimposed

Chapter 4

The Quantitative Origin of the Digital-Numerical Phenomenon

Benford's Law is stated purely in terms of the proportions of the digits within numbers in data sets, and as such it is based on and depends on our number system. Indeed, there exists a (misguided and mistaken) school of thought that holds that Benford's Law is merely a consequence of our arbitrarily invented number system and that it is a numerical phenomenon, not a physical phenomenon. In other words, it holds that the law reflects the aspects and features of our positional number system and that it is not a law of nature expressing physical and natural reality. As shall be discussed in the last section of this book, this vista of Benford's Law is absolutely wrong.

When the conceptual distinctions between digits, numbers, and quantities are explored, it leads to the key finding that the phenomenon is actually quantitative in nature, originating from the fact that in extreme generality, nature creates many small quantities but only very few big quantities, corroborating the motto "small is beautiful". Indeed, histograms typically fall on the right, that is, they are nearly always positively skewed and are not symmetrical. Real-life data histograms very rarely rise or remain flat overall in the aggregate from the minimum to the maximum. The Normal distribution constitutes but a tiny minority in the universe of data, in spite of being essential and occurring in some very important scenarios and physical phenomena, while the Uniform distribution is even rarer and it almost never manifests itself in real-world data.

Furthermore, let us examine the relationship between some particular sub-intervals on the positive x-axis and the first-digit occurrences of numbers residing within them:

- On the sub-interval $(1, 2)$, with typical numbers such as 1.24, 1.65, and 1.97, digit 1 leads.
- On the sub-interval $(10, 20)$, with typical numbers such as 11.6, 15.9, and 18.3, digit 1 leads.
- On the sub-interval $(100, 200)$, with typical numbers such as 128, 151, and 194, digit 1 leads.

- On the sub-interval $(2, 3)$, with typical numbers such as 2.35, 2.74, and 2.92, digit 2 leads.
- On the sub-interval $(20, 30)$, with typical numbers such as 21.8, 23.2, and 28.6, digit 2 leads.
- On the sub-interval $(200, 300)$, with typical numbers such as 215, 249, and 293, digit 2 leads.

Figure 8 helps in visualizing the relationship between the x-axis locations and first-digit occurrences.

If the histogram is positively skewed in the aggregate, mostly falling to the right, then in extreme generality, the three sub-intervals $(1, 2)$, $(10, 20)$, and $(100, 200)$, where first digit 1 dominates, are of higher frequency and stronger density than the three adjacent sub-intervals just to the right of each of them, namely $(2, 3)$, $(20, 30)$, and $(200, 300)$, where first digit 2 dominates, hence digit 1 occurs more frequently than digit 2 in the first place of numbers. The same argument is applied to any other pair of neighboring digits; for example, a higher and taller curve for digit 5 versus a lower and shorter curve for digit 6, demonstrating that higher digits nearly always obtain fewer data points than do lower digits when histograms fall to the right in the aggregate. Figure 8 depicts this histogram-digital rule, assuming positive skewness. Clearly, the quantitative phenomenon causes and drives the digital phenomenon. Small quantities outnumber big quantities; consequently, low digits outnumber high digits. Specifying and defining the histogram decisively determines the digital structure of the data set for which the histogram is drawn. But specifying and defining the digital structure of a given data set does not determine any unique histogram.

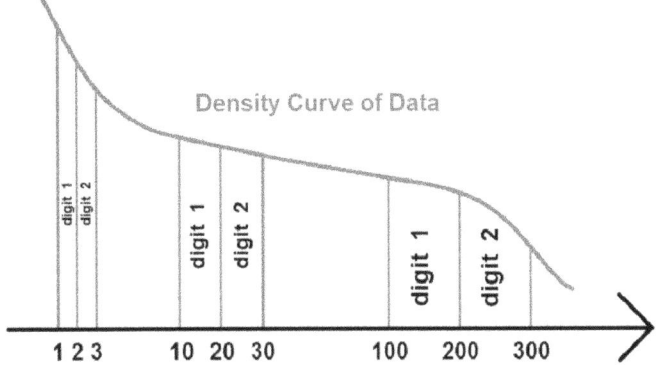

Figure 8. Falling Data Curve Allocates More Areas to Lower Digits

To recapitulate, if the histogram is positively skewed in the aggregate and mostly falling to the right overall from the minimum to the maximum globally, then necessarily it is also falling to the right locally, on all or on the majority of the sub-intervals $(1, 10)$, $(10, 100)$, $(100, 1000)$, and so forth, where on each of them the first digits complete a full cycle from 1 to 9. All or most of the parts of the histogram hovering over these crucial first-digit sub-intervals are declining (sharply or gently); as a consequence, the net result is a strong preference for low digits, as the quantitative phenomenon of positive skewness with a tail falling to the right produces and leads to the digital phenomenon of Benford's Law.

This is also why the second digits are more evenly balanced, with only mild differences in probabilities. The cycles of the second digits are much shorter, occurring on much narrower sub-intervals. For example, on the narrow sub-interval of $(1.0, 2.0)$, we progress and rotate from 0 as the second digit to 9 as the second digit, as in, say, 1.0, 1.2, 1.3, 1.4, 1.5, 1.6, 1.7, 1.8, 1.9. Other examples of complete second-digit cycles are $(2.0, 3.0)$, $(3.0, 4.0)$, $(10, 20)$, $(20, 30)$, $(100, 200)$, $(200, 300)$, and so forth, where a full second-digit cycle occurs on each of these shorter sub-intervals, rotating from 0 to 9. Therefore, *locally*, from the perspective of the second digits and their much narrower span, the fall in the histogram is certainly less pronounced and milder compared to the *global* and dramatic fall in the histogram from the perspective of the first digits with a much wider span. For the third digits, having even shorter cycles, such as (1.00, 1.10), (1.10, 1.20), (1.20, 1.30), (10, 11), (11, 12), (100, 110), (110, 120), and so forth, the *local histogram* appears almost horizontal; therefore, the third digits attain near equality of occurrences. My home garden and even my entire city appear flat, even though I live on a round, three-dimensional sphere called Earth, but the whole Asian continent or the whole European continent appears round, as does planet Earth. Astute engineers might be able to measure the gentle curve of my three-dimensional city, but in reality the city is nearly flat, even though Earth is not flat.

Indirectly and subtly, by examining digital configurations, Benford's Law in essence numerically quantifies the "small is beautiful" phenomenon and obtains an exact measure indicating how much the relatively small is more numerous than the relatively big.

The discoverer himself, Frank Benford, believed it was a physical phenomenon, eloquently writing in his seminal 1938 article his philosophical approach to the phenomenon: "As has been pointed out before, the theory of anomalous numbers [Benford's Law] is really the theory of phenomena and events, and the numbers but play the poor part of lifeless symbols for living things".

The generic causes leading to the "small is beautiful" phenomenon (such as multiplication processes, partition processes, some consolidation processes, data aggregation, and data dependency) typically work gradually or incrementally, possibly starting with Normal, Uniform, and other non-skewed symmetric types of raw data, then leading to mild positive skewness, stopping well short of the decisively skewed Benford configuration. Typically, more vigorous or simply repeated applications of these causes converge and lead fully to the decisively skewed Benford configuration. Hence, the "small is beautiful" phenomenon has a much wider scope and is much more prevalent in the physical world, in factual data, and even in series, distributions, and other entities in the realm of abstract mathematics than the more particular Benford quantitative configuration. This statement does not imply that Benford's Law is not prevalent in scientific, physical, and numerous other data types; on the contrary, it is highly prevalent. The statement only implies that in almost all the counterexamples and exceptions to Benford's Law, the "small is beautiful" phenomenon is still valid, albeit with different quantitative configurations than that of the Benford type and typically milder and less skewed, but at times with even more skewness.

To emphasize, all these points can be stated more succinctly in three ways:

- **(I)** Benford's Law configuration is a subset of the "small is beautiful" phenomenon.
- **(II)** The "small is beautiful" phenomenon is even more prevalent than Benford's Law.
- **(III)** A significant portion of non-Benford real-life data is quantitatively structured in the spirit of the "small is beautiful" phenomenon.

Chapter 5

The Scale Invariance Principle

A change in unit or scale profoundly affects the appearance and digital structure of any given data set, yet in spite of this, as if by magic, Benford's Law still holds! The lengths of the immense rivers {Nile, Amazon, Yangtze, Mississippi, Congo, Volga} in units of kilometers are {6,650, 6,400, 6,300, 6,275, 4,700, 3,645}, but in units of miles the lengths are {4,135, 3,967, 3,917, 3,902, 2,922, 2,266}. Clearly, the first, second, third, and fourth digits here all change completely. Yet, the data set regarding the lengths of all the rivers worldwide is Benford for all digital orders, and this is so when measured either in kilometers, in miles, or in any other length unit. One mile is equivalent to 1.609 kilometers. The multiplication of all river length values expressed in miles by a singular conversion factor of 1.609 in order to express values in kilometers has almost no effect on the resultant proportions of the first, second, third, and fourth digits, and compliance with Benford is maintained under such scale conversion.

Benford's Law is valid assuming the use of any scale whatsoever in measuring a physical phenomenon and generating the data set under consideration. Surely, there is nothing special about kilograms, meters, feet, seconds, hours, inches, kilometers, or miles; they are all arbitrary, and philosophically, for the law to be considered consistent, it should hold true in any units and scales. Had Benford's Law been valid say for the data set on the length of all rivers worldwide measured in kilometers but totally false if that length was measured in miles, then trust and appreciation of the law would have been lost, and the digital status of data sets regarding the length of other physical entities in the world would have been in doubt, resulting in total confusion with regard to the question of what "proper" unit should be used for the law to be valid. The fact that Benford's Law is indeed independent of the societal scale system and the arbitrary units in use renders it universal, and it even endows it some mystique. This property of the law is called "The Scale Invariance Principle".

As an example, Figure 9 depicts the results of the first digits — on various scales — for the data on the time between all successive earthquakes worldwide for the year 2012, totaling 19,452 earthquakes. In other words, what is measured here is the length of time between any one earthquake occurring somewhere in the world and the next global earthquake occurring immediately after it. During the year 2012, on average, an earthquake occurred somewhere in the world every 27 minutes, but they were mostly very mild earthquakes causing no damage whatsoever and almost unnoticeable. The digital results for the first digits shown in Figure 9 are all very near the Benford proportions, regardless of the choice of time scale in use. The second and third digits here for the earthquake data are not shown, but they are also Benford, regardless of scale, with only minor deviations due to scale changes.

Data Set	1	2	3	4	5	6	7	8	9
Earthquake (Second)	29.9	18.8	13.5	9.3	7.5	6.2	5.8	4.8	4.2
Earthquake (Minute)	28.6	17.7	12.8	10.5	8.6	6.8	5.8	5.0	4.2
Earthquake (Hour)	29.2	16.9	12.3	9.8	8.1	7.0	6.2	5.7	4.9
Earthquake (Day)	29.4	18.2	13.6	10.2	7.6	6.3	5.4	4.9	4.5

Figure 9. Benford is Confirmed in Various Scales for Earthquake Data

The small deviations seen here between these four scales are due to the fact that the earthquake data set is not as perfectly Benford as it could be, in spite of being very close to it, and also because it has only a finite number of data points. Theoretically, had it been perfectly and exactly Benford with an infinite number of data points, then there would have been no deviations at all due to scale changes. In practical terms, had it been extremely close to Benford and nearly perfectly so, with a truly colossal number of data points, then there would have been only minuscule and barely noticeable deviations due to scale changes. It should be noted that a scale change, say, from minutes to seconds, entails multiplying each time interval quoted in minutes by a factor of 60. Hence, the more general view of the scale invariance principle is that the digit distribution of Benford-type data remains nearly unchanged under a multiplicative transformation of all the data points by an identical multiplicative factor. It should also be noted that the principle does not imply that the digital structure of non-Benford data sets is constant under a scale change. It is not! Only Benford data sets are endowed with the constancy of their Benfordian digital structure under a scale change.

Chapter 6

The Base Invariance Principle

Benford's Law is valid for all positional number systems of whatever base B, including of course our decimal base 10 system. This important generalization is accomplished by using the appropriate logarithm base B in the expression of the probabilities for the first leading digit d, varying in the range from 1 to $B - 1$. The generalized Benford's Law for any base B is

Probability[First Leading Digit is d] $= \text{LOG}_B(1 + 1/d)$

For a positional number system of base 6 for example, with {1, 2, 3, 4, 5} as the set of all possible first digits, Benford's Law predicts the probabilities of {38.7%, 22.6%, 16.1%, 12.5%, 10.2%}, calculated as {$\text{LOG}_6(1 + 1/1)$, $\text{LOG}_6(1 + 1/2)$, $\text{LOG}_6(1 + 1/3)$, $\text{LOG}_6(1 + 1/4)$, $\text{LOG}_6(1 + 1/5)$}.

Our choice of base 10 for our number system is a natural one, having 10 moveable fingers. We literally count with the use of our fingers for small numbers as kids, and even as adults. Yet, the cosmic perspective is that base 10 is an arbitrary choice. The fact that Benford's Law is indeed independent of the choice of the base in a positional number system renders it universal and consistent. This property of the law is called "The Base Invariance Principle".

Chapter 7

Physical Order of Magnitude of Data

Order of magnitude (OOM), namely variability of data, plays a crucial role in the context of Benford's Law, since the law is valid (with only one rare exception) for data sets with a high order of magnitude, namely for data of sufficient variability. The Physical Order of Magnitude (POM) of a given data set is an alternative measure expressing the extent of data variability. It is defined as the ratio of the maximum value to the minimum value. The data set is assumed to contain only positive numbers greater than zero.

$$POM = Maximum/Minimum$$

The classic or orthodox definition of order of magnitude involves the application of the logarithm to the ratio maximum/minimum, transforming it into a smaller and more manageable number.

$$OOM = LOG_{10}(Maximum/Minimum)$$
$$OOM = LOG_{10}(Maximum) - LOG_{10}(Minimum)$$

Visual perception of classic OOM is quite easily obtained by simply examining the span on the log-axis of the histogram of decimal base 10 logarithmically transformed values, namely the distance on the log-axis from its minimum log value to its maximum log value — excluding whiskers and isolated or remote outliers, perhaps for the sake of robustness. The avoidance of logarithm in the POM definition is motivated by the fact that such a logarithmic transformation has a monotonic one-to-one relationship with max/min, hence it does not provide for any new insight or information, but could rather be looked upon in a sense as simply the use of an alternative unit or different scale, still measuring the same thing. For this reason, the complexity of the logarithm can be avoided altogether by referring only to the simple POM measure. The more profound reason for using POM instead of OOM is its feature as a universal measure of variability, totally independent of the particular societal number system in use, as well as being independent of the arbitrary choice of base 10. This is the motivation behind the use of the term "physical", alluding to something real, natural, and existing in its own right.

Since both OOM and POM are defined using the ratio of two quantities, they are unitless and independent of the particular societal scale used to count or measure the physical quantity. If the ratio of the maximum of 360 kilograms weight to the minimum of 30 kilograms weight is 360/30, namely 12, then such a measure of variability, expressed as 12 or $\text{LOG}_{10}(12)$, is conserved under its transformation to pound, gram, or ton units. Let it be emphasized once again: The OOM and POM measures of variability come with a touch of universality, as they are scale invariant, true under the use of whatever arbitrary units are chosen to measure physical reality. With regard to compliance with Benford's Law, the only variability measure that matters is the scale-invariant OOM or POM, while the standard deviation is totally irrelevant. Standard deviation depends on the arbitrary societal definition of scales and units, lacking universality. Since Benford's Law itself is scale and base invariant, it is altogether fitting and proper that we construct a measure for its compliance that is also scale and base invariant.

The points 0.001, 0.01, 0.1, 1, 10, 100, 1000, 10000, 100000, and so forth are often used as important marks along the x-axis for the raw data itself to suggest or approximate the order of magnitude of the data. Between any two such consecutive points, such as 10 and 100, the order of magnitude is 1; between 10 and 1000, the order of magnitude is 2; and so forth. Hence one advantage of OOM is that it signifies roughly the number of whole first digit cycles the data goes through. For example, if the minimum is 11 and the maximum is 9972, then OOM is about 3, and the data indeed spans nearly 3 full first digit cycles.

Chapter 8

Robust Measure of Physical Order of Magnitude

A robust definition of OOM should prove steady and consistent, strongly resisting outliers and preventing them from overly influencing or exaggerating the numerical measure of data variability. For example, the data set $\{2, 25, 26, 27, 28, 29, 32, 33, 34, 36, 37,$ $39, 40, 41, 42, 47, 48, 50, 53, 55, 56, 57, 59, 60, 63, 67, 68, 75,$ $77, 78, 79, 80, 84, 86, 91, 94, 103, 112, 119, 315, 738\}$ has three apparent outliers, namely $\{2, 315, 738\}$. Simplistic calculations lead to OOM $=$ $\text{LOG}_{10}(738/2) = \text{LOG}_{10}(369) = 2.6$, but this is an exaggeration. With the elimination of these three obvious outliers, a more realistic OOM measure is calculated as $\text{LOG}_{10}(119/25) =$ $\text{LOG}_{10}(4.8) = 0.7$, which nicely explains the decisive non-Benford behavior of this data set. In other words, these three outliers inflate and exaggerate the measure of variability, and only once we eliminate them can we measure variability in a robust and realistic way.

Most outliers in real-life data are not as visually obvious as in the above data example, hence the need for a systematic, mechanical, and automated way of going about eliminating outliers. This is accomplished by narrowing the focus exclusively onto the core 98% of the data, excluding the top 1% and the bottom 1%, which are considered the "edges" of the data. This brutal purge eliminates without mercy any malicious or misleading outliers, as well as any innocent and proper data points which just happened to stray a little bit away from the core part of the data. The measure is called Core Physical Order of Magnitude (CPOM) and is defined as follows:

$$CPOM = P_{99\%}/P_{1\%}$$

The definition simply reformulates POM by substituting the 1st percentile (in symbols, $P_{1\%}$) for the minimum and by substituting the 99th percentile (in symbols, $P_{99\%}$) for the maximum.

The definition of percentile requires us initially to order all data points from the minimum to the maximum. The 1st percentile is the value below which about 1% of the data points may be found. The 50th percentile is the median, below which about half of all the data points may be found. The 99th percentile is the value below which about 99% of the data points may be found.

For the above small data example, $LOG_{10}(CPOM) = LOG_{10}(P_{99\%}/P_{1\%}) = LOG_{10}(568.8/11.2) = LOG_{10}(50.8) = 1.7$, constituting a more realistic and robust order of magnitude.

For a stricter procedure, with stronger emphasis on eliminating all possible outliers but less emphasis on the avoidance of losing authentic data points, the narrower range from the 5th percentile to the 95th percentile might be considered, applying the ratio $P_{95\%}/P_{5\%}$ for a measure of the core 90% of the data. At times, when a very strong suspicion about widespread and tricky outliers exists, a much stricter procedure is applied by considering the even smaller range from the 10th percentile to the 90th percentile and applying the ratio $P_{90\%}/P_{10\%}$ for a measure of the core 80% of the data.

These two strict approaches (core 90% and core 80%) should be considered a bit extreme and used predominantly for problematic or odd data sets. This is akin to a surgeon's dilemma of how much to cut around the malignant tissue in a skin cancer removal surgery. The responsible doctor should also cut around the tumor itself, even into healthy tissue, in order to ensure that all malignant cells (outliers) are gone and that the cancer does not return. This is more of an art than a science, more subjective than objective, and even philosophical. The default preference of this author, following extensive empirical data observations in the most generic way for many years in the context of Benford's Law, is to use the 1st and 99th percentiles as the standard robust measure, containing the core 98% of data.

Chapter 9

Two Essential Requirements for Benford Behavior

One of two essential prerequisites or conditions for data configuration with regard to compliance with Benford's Law is that the value of the OOM of the data set should be over 3 approximately, namely $LOG_{10}(Maximum/Minimum) > 3$. This implies that $(Maximum/Minimum) > 10^3 = 1000$. Therefore, the threshold rule for the POM value separating compliance from non-compliance is approximately POM > 1000, constituting the condition for compliance.

The above prerequisite for compliance totally ignores the thorny issue of outliers and edges, and in that sense it is too simplistic and even completely erroneous for numerous types of data sets. Hence, using the more robust CPOM qualification is essential in judging whether or not a given data set is expected to comply with Benford's Law. The proper qualification for expectance of compliance with the law in the approximate — obtained via extensive empirical studies — is then as follows:

$$CPOM = P_{99\%}/P_{1\%} > 1000$$

This requirement is very general as it is valid with respect to any generic B base number system in addition to our decimal base 10 number system, so that the inequality is always stated as such without referring to any logarithm or any particular base B. If one insists on OOM and logarithmic narratives, then this requirement can be stated in any Base B as:

Min Core $OOM(P_{99\%}/P_{1\%})$ = $LOG_{BASE}(P_{99\%}/P_{1\%})$ > LOG_{BASE}(one thousand)

For example in hexadecimal positional number system with Base 16:

Min Core $OOM(P_{99\%}/P_{1\%})$ = $LOG_{16}(P_{99\%}/P_{1\%})$ > $LOG_{16}(3E8) = 2.7D70A3D$

In positional number system with Base 7:

Min Core $OOM(P_{99\%}/P_{1\%})$ = $LOG_7(P_{99\%}/P_{1\%})$ > $LOG_7(2626) = 3.35641620521$

Hence while in decimal Base 10 the LOG of the ratio must be at least 3.0, in Base 16 however it must be at least 2.5, and in Base 7 it must be at least 3.5 (and these three thresholds here in the last sentence are written via our own decimal number system - for a good comparison).

Actually, even a lower CPOM value of, say, 600 is expected to yield the Benford configuration approximately, but falling below 300 does not bode well for getting anywhere near the Benford distribution. As it happens, nearly all real-life physical data sets, as well as financial, accounting, informational, and census data, are generally of sufficient variability, where the CPOM value is definitely over 1000 or at least over 600, and consequently Benford's Law nearly always prevails. Most big-size data sets of scientific and natural phenomena have a CPOM value of over 5000 and even over 10000!

Skewness of data, where the histogram comes with a prominent tail falling to the right, is the second essential criterion necessary for Benford behavior. Indeed, nearly all real-life physical and scientific data sets are generally positively skewed, and consequently the quantitative configuration is such that the small is numerous and the big is rare, implying that low digits decisively outnumber high digits.

The *asymmetrical* Exponential, Lognormal, and k/x distribution are typical examples of such quantitatively skewed configurations; therefore, they are approximately, nearly, or exactly Benford, respectively. The *symmetrical* Uniform, Normal, evenly triangular, circular-like, and other such distributions are inherently non-Benford, or rather anti-Benford, as they lack skewness and do not exhibit any bias or preference toward the small and the low.

Symmetrical distributions are always non-Benford, no matter what values are assigned to their parameters. By definition, they lack that asymmetrical tail falling to the right, and such a lack of skewness precludes Benford behavior, regardless of the value of their OOM. The OOM simply does not play any role whatsoever in Benford behavior for symmetrical distributions. For example, the Normal(10^{35}, 10^{8}) and Uniform(1, 10^{27}) distributions are not Benford at all, and this is so in spite of their extremely large OOMs.

In summary, Benford behavior in extreme generality can be found with the confluence of a sufficiently large OOM together with a positive skewness of data. The combination of positive skewness and large OOM is not a guarantee for Benford behavior, but it is a strong indication of likely Benford behavior under the right conditions. Modest (overall) quantitative skewness with a tail falling too gently to the right implies that digits are not as skewed as in the Benford configuration. Extreme (overall) quantitative skewness with a tail falling too sharply to the right implies that digits are severely skewed, even more so than they are in the Benford configuration.

An informal and associated addendum or extension to Benford's Law indirectly hints that the vast majority of real-life data sets are indeed positively skewed and come with a high OOM. Benford's Law also indirectly states that the overall or aggregate quantitative skewness is of the Benfordian Log(1 + 1/d) manner, that the fall of the histogram on the right is neither too modest and gentle nor too sharp and steep, and that it falls at an intermediary rate somewhere between these two extreme cases. The only one rare exception to the generic rule of high OOM for Benford behavior is the perfectly Benfordian k/x distribution defined over two adjacent integral powers of 10, such as (1, 10) or (10, 100), having a very low OOM value of 1.0, as shall be discussed in later chapters.

Chapter 10

Sum of Squared Deviations Measure

It is necessary to establish a standard measure of "distance" from the Benford digital configuration to the particular digital configuration of any given data set. Such a numerical measure might tell us about the degree of closeness of the data set under consideration to Benford. This is accomplished with what is termed **sum of squared deviations (SSD)**, defined as the sum of the squares of the nine "errors" between the Benford expectations and the observed values. The definition uses percentage values, as opposed to fractional, proportional, or probability format, but without the "%" sign, such as 30.1 for Benford digit 1 theoretical expectation.

$$\text{SSD} = \sum (\text{observed \% of digit } d - 100 \times \text{LOG}(1 + 1/d))^{\mathbf{2}}$$

with summation index d running from 1 to 9 for the first-digit law. For example, for the observed percentages of the first digits for a given data set, as in {31.1, 18.2, 13.3, 9.4, 7.2, 6.3, 5.9, 4.5, 4.1} with the "%" sign omitted, the SSD measure of distance from Benford is calculated simply as

$$
\begin{aligned}
\mathbf{SSD} = {} & (31.1 - \mathbf{30.1})^2 + (18.2 - \mathbf{17.6})^2 + (13.3 - \mathbf{12.5})^2 \\
& + (9.4 - \mathbf{9.7})^2 + (7.2 - \mathbf{7.9})^2 \\
& + (6.3 - \mathbf{6.7})^2 + (5.9 - \mathbf{5.8})^2 \\
& + (4.5 - \mathbf{5.1})^2 + (4.1 - \mathbf{4.6})^2 = \mathbf{3.4}
\end{aligned}
$$

The SSD measure is also applied for higher-order digit distributions, digital combination distributions, and distributions within other bases in positional number systems, as follows:

$$\text{SSD} = \sum (\text{Observed } \% - \text{Theoretical } \%)^2$$

with summation index running from: 0 to 9 for the second, third, and higher digital orders in our decimal number system; 10 to 99 for the first two-digits combinations (in Chapter 39); 00 to 99 for the last two-digits combinations (in Chapter 39); and (1) to $(B - 1)$ for the first digits in another base B number system.

The fractional or proportional format of deviations, such as $(0.301 - 0.285)$, are typically of very small fractional values when squared, making it a bit difficult to internalize, memorize, and compare empirical results with threshold values. The percentage format of deviations, such as $(30.1 - 28.5)$, on the other hand, yields typically larger values when squared, which are easier to work with. This is the motivation behind using this format for the SSD definition.

Since SSD does not incorporate in its expression the term for the data size (denoted by N), one gets the same measure and conclusion regardless of the number of data points in the data set. The unimportant and insignificant price to pay for omitting N here is that there can be no associated statistical theory to guide us (supposedly) in calculating objectively a definite limit or threshold above which researchers could conclude with, say, a 5% confidence level that a given data set is deviating from Benford's Law. Surely, if our data set could somehow be depicted as a truly random sample from some supposedly much larger Benford or non-Benford parental universe, then statistical theory cannot be indifferent to data size N, since without knowing how many values have been collected as samples from that supposedly parental universe, there is no way to tell whether a certain deviation from Benford is due to chance (and that the parental universe is actually Benford) or due to structural causes (with the parental universe being indeed non-Benford). While N is absent in the expression of SSD, and if the data is somewhat different from Benford, we do not know whether this deviation from Benford is due to chance (i.e., not an unlikely nor very rare sampling result, so it does not reflect strongly on the parental universe, which might still be Benford) or due to structural causes (i.e., a decisive sampling result, which does necessarily reflect on the parental universe, and thus the parental universe is now definitely known to be non-Benford).

The law of large numbers, in probability and statistics, states that as the sample size N grows, its average gets closer to the average of the whole population. This is due to the sample being more representative of the whole population as the sample size N becomes larger. If the lucky side of six appears only once in 20 casino dice throws, the casino might be honest, but if the lucky side of six appears only once in 100 dice throws, the casino might be declared fraudulent. In reality, no such statistical framework or basis exists for nearly all real-life data types anyhow, since almost all of them are not such that individual points are methodically "collected" or "selected" from any (imaginary) larger parental pool of numbers or universe, one data point at a time, randomly and independently of each other. Therefore, the price paid for omitting N is totally insignificant. SSD is therefore applied only as a measure of distance from Benford, and one has to subjectively judge a given SSD value of the data on hand to be either low enough and thus somewhat close to Benford or too high and thus definitely non-Benford in nature.

This pessimistic conclusion and the challenging state of affairs regarding compliance with Benford's Law cannot be changed even with the incorporation of N somewhere within the SSD expression in some statistically clever way or format. Attempts to attribute such a statistical random selection framework to any real-life data type are almost always nothing but an illusion and wishful thinking. A systematic yet non-statistical, non-theoretical way of going about establishing practical SSD cutoff points or threshold values is to empirically compare the SSD value of the data set under consideration with a big list of SSD values of a large variety of honest and real data sets. This author has performed exactly such inter-data generic comparisons and observations in the most reasonable way and as evenhandedly as humanly possible perhaps. Such accumulated knowledge helps us decide the significance of those SSD values.

The author's guide for the first digits, second digits, first-two-digits combinations, and last-two-digits combinations is that an SSD reading below 2 should be considered ideally Benford, and that an SSD should be below 25 for the data to be accepted as generally Benford; but any data with an SSD over 100 are considered to deviate too much from Benford. Actually, for the second digits and first-two-digits combinations, perhaps a stricter upper limit of 50 is warranted, beyond which the data should be considered as deviating too much from Benford.

This SSD measure does not answer the question: "Does the data obey Benford's Law or not?" nor the question: "Is the (tiny, mild, small) digital deviation from Benford for the given data set due merely to chance and randomness or is it structural and intrinsic?" This SSD measure does not respond to any "yes or no" types of questions. Rather, SSD answers the question: "How far is the data set from Benford?" or: "How close is the data set to Benford?"

A famous alternative to the SSD measure is called "Mean Absolute Deviations", abbreviated as MAD for its acronym, and defined as the sum of all the absolute value deviations, divided by the number of deviations – namely divided by the number of digits involved. This author choice of SSD instead of MAD was made for two reasons. The first reason is that MAD numbers are typically of very small fractional values which are difficult to memorize or internalize intuitively. The SSD measure on the other hand yields typically much larger values and thus it is easier to work with. But the more serious theoretical reason for rejecting MAD is its division of the total absolute deviations by the number of digits involved, and which gives a wrong measure of deviation. The focus should be on "the entire deviation of the data set from Benford", instead of the misguided focus on "deviation from Benford per digit involved". This not-so-subtle difference in conceptual approach becomes very obvious (and disturbing) when dealing with a wide variety of B Bases in our positional number system, where for truly Benford data with a strong OOM, SSD is steady and consistent for all B base choices, while MAD values decrease rapidly and superficially for large B bases, as shall be seen at the end of Chapter 38. The misguided motivation of the MAD approach is to counter the increasing numbers of differences between the theoretical and the empirical as the number of digits becomes large, thus perceiving that to cause total sum of absolute differences to increase, leading to the perceived need to oppose this supposed increase by dividing by the number of digits. But in reality, total sum does not increase as the number of digits increases, because that fixed 100% total of theoretical or empirical is being spread, divided, and shared, among many more digits, getting diluted, each digit obtaining an even smaller share of the 100% total, thus automatically causing each difference to be of a lesser value. In other words, we do not need to proactively correct anything here, because the measurement (SSD or total absolute differences) self-adjusts by default.

Chapter 11

The Mistaken Use of the Chi-Square Test in Benford's Law*

The definition of the chi-square statistic and its related test are as follows:

$$\text{chi-square} = (N) \times \sum \left((\text{Observed} - \text{Theoretical})^2 / \text{Theoretical} \right)$$

"Observed" and "Theoretical" are in fractional, proportional, and probability format.

Reject the null hypothesis H_0 that the data set is authentically Benford at the $p\%$ confidence level if the chi-square statistic in the expression above is larger than the chi-square $p\%$ critical (threshold) value with eight degrees of freedom for the first digits (namely, reject if it's over 15.5 if $p\%$ is chosen as 5%); and reject it if the statistic is larger than the chi-square $p\%$ critical (threshold) value with nine degrees of freedom for the second digits (namely, reject if it's over 16.9 if $p\%$ is chosen as 5%).

Readers are strongly advised to completely avoid the chi-square test. Nearly always, the use of the chi-square test in the context of Benford's Law is without proper mathematical-statistical basis and thus erroneous. Yet, due to groupthink influences and behavior,

---

*A detailed discussion about this issue can be found in an article by the author on **www.mdpi.com**, titled "On the Mistaken Use of the Chi-Square Test in Benford's Law", as well as in Section 10 of the author's book, *Studies in Benford's Law*.

researchers, data analysts, and statisticians typically rush to apply the chi-square test in order to attempt to verify compliance or deviation from Benford, and this perhaps might give the illusive satisfaction of being assertive and decisive, but it is wrong — so much so that the chi-square test has become a dogma, or rather an impulsive ritual, in the field of Benford's Law, being applied to whatever data set the researcher is considering, regardless of its true applicability. The mistaken use of the chi-square test has led to much confusion and many errors and has considerably undermined trust and confidence in the whole discipline of Benford's Law. There is a need to correct course and bring rationality and order to this field, which had demonstrated harmony, beauty, and consistency in all of its results, manifestations, and explanations.

Since the chi-square test incorporates the term N, it implies that whenever the size of the data set is quite large, even a seemingly mild deviation from Benford can still result in a fairly large value of the chi-square statistic, thus indicating non-compliance with Benford's Law. Because of this fact, the chi-square test is mistakenly thought to be "oversensitive" in the sense that, for large data sets, supposedly even mild deviations from Benford are flagged as significant ("false positives"). Also, whenever the data size N is very small, the chi-square test suddenly appears to be very liberal and too permissive. Thus, the test is mistakenly thought to be "undersensitive" in the sense that, for small data sets, supposedly even large deviations from Benford are tolerated as insignificant ("false negatives"). These misguided perceptions of the chi-square test are simply an indication that users lack understanding of the underlying statistical basis of the test.

In essence, the error of the chi-square application originates from the fact that real-life data sets almost never arise from random and independent selections of data points from some much larger (imaginary) universe of parental data, one data point at a time and independently of each other, as the chi-square approach supposes. For example, the population value of Chicago in the 2018 US Census data cannot be described as a purely random pick of a single and simple number from some enormous or infinite pool of galactic population numbers, which may or may not be authentically Benford. Rather, the current population of Chicago is the cumulative probabilistic result of numerous deterministic factors occurring throughout the past centuries, relating to birth and death rates, weather, climate, agricultural history, health practices, medical knowledge, immigration and migration to and from adjacent cities and other far-away continents, as well as other issues influencing population. This exact population number for Chicago is surely (at a minimum) not independent of the population values of its neighboring cities, due to constant immigration and migration. The collective set of all the many thousands of values regarding all population centers in the USA for all its cities and towns, is indeed the entire data set, standing distinctly and independently and existing in its own right, as opposed to a collection of random selections of numbers from some imaginary and non-existent larger population data pool.

SECTION II
Causes and Explanations

Chapter 12

Multiplication Processes Lead to Positive Skewness and Often to Benford

Almost all random and deterministic multiplication processes induce a dramatic increase in skewness, where the small becomes relatively numerous and the big becomes relatively rare. Multiplication processes also induce a decisive increase in order of magnitude, albeit gradually, increasing steadily with each extra multiplicative step, until it finally leads to Benford behavior whenever OOM reaches the threshold of about 3. Once the skewed Benford configuration is attained and OOM is 3, saturation sets in, and any additional multiplications beyond that point do not lead to any increase in digital or quantitative skewness, yet OOM continues to increase indefinitely.

Let us examine the quantities within the 10-by-10 multiplication table, which we all were forced to memorize during our elementary school years against our will. In this example, the intrinsic characteristics of multiplication processes with regard to quantitative resultant sizes and first digits shall be explored. Such an analysis is done at the most primitive and basic level and at the arithmetic and quantitative level, before going on to any rigorous mathematical level. The quest is to start out with this very particular example and then extrapolate it, lending the conclusions derived from this case universality and applicability in almost all multiplication processes.

The entire range from 1 to 100 is to be partitioned fairly into 10 quantitative sections of an equal 10-unit width each, namely into [1, 10], [11, 20], [21, 30], [31, 40], [41, 50], [51, 60], [61, 70], [71, 80], [81, 90], and [91, 100], where the sections are initially empty, and then we let the values from the multiplication table fall carefully within each section according to their size. The goal is to group the 100 products of the multiplication table according to their sizes. Figure 10 depicts this quantitative partitioning arrangement of the entire multiplicative territory by size. Figure 11 depicts the histogram of the values falling within each section. Surprisingly, a decisive trend regarding the occurrence of products within the sections is found here.

The sequence of the number of values falling within these 10 territorial sections and presented in order according to their size is {27, 19, 15, 11, 9, 6, 5, 4, 3, 1}. Clearly, the sections pertaining

X	1	2	3	4	5	6	7	8	9	10
1	1	2	3	4	5	6	7	8	9	10
2	2	4	6	8	10	12	14	16	18	20
3	3	6	9	12	15	18	21	24	27	30
4	4	8	12	16	20	24	28	32	36	40
5	5	10	15	20	25	30	35	40	45	50
6	6	12	18	24	30	36	42	48	54	60
7	7	14	21	28	35	42	49	56	63	70
8	8	16	24	32	40	48	56	64	72	80
9	9	18	27	36	45	54	63	72	81	90
10	10	20	30	40	50	60	70	80	90	100

Figure 10.   Quantitative Territorial Partitioning of the Multiplication Table

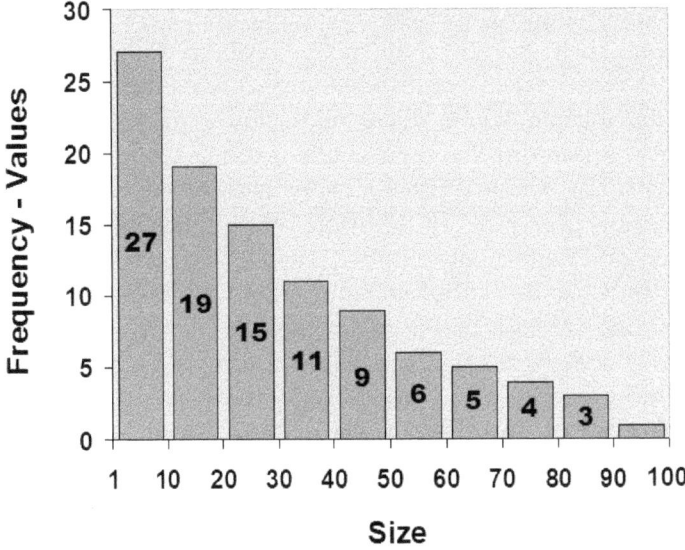

Figure 11. Skewed Histogram of the 10-by-10 Multiplication Table

to bigger quantities have fewer values falling within them, as the count of values monotonically decreases. The small is definitely more numerous than the big in our standard childhood multiplication table. The crucial lesson learned from this multiplicative process is that this tendency is nearly universal, and that it should be present in almost all other multiplication processes, not only for our particular 10-by-10 multiplication table. There is nothing unique about our standard multiplication table or the set of numbers from 1 to 10.

Examining proportions of the first digits instead of quantities, it is noted that out of 100 numbers, 21 numbers start with the digit one, while only 5 numbers start with the digit nine. The vector count of all the nine first digits is {21, 17, 13, 14, 8, 9, 6, 7, 5}, which is also the percent vector here since we divide counts by 100 values total. The above percent vector of the first digits of the multiplication table is a bit similar to Benford's Law! Its SSD value is 111.2. In order to get much closer to the Benford configuration, either a larger set of numbers is multiplied, say from 1 to 1000, instead of 1 to 10, or else the set of 1 to 10 is multiplied by itself three or four times, instead of merely twice. Surely, our multiplication table could also be interpreted as the multiplicative process of two continuous Uniform distributions, namely the product of Uniform$(1, 11) \times$ Uniform$(1, 11)$. Hence, from such an even, uniform, and balanced distribution of the Uniform, we arrived via multiplications at a decisively skewed and uneven distribution, where the small is more numerous than the big, resulting in a Benford-like digit configuration.

Conceptually, the quantitative effects drive the digital effects in multiplication processes. In addition, these quantitative tendencies of increasing skewness correlate with a decisive increase in the order of magnitude of the resultant multiplicative data. Our 10-by-10 multiplication table is based on two discrete Uniform distributions $\{1, 2, 3, 4, 5, 6, 7, 8, 9, 10\}$, each with only a Log$(10/1)$ order of magnitude, namely 1, and its multiplicative action leads to a set of numbers with a Log$(100/1)$ order of magnitude, namely 2, thus doubling the order of magnitude.

In general, random and deterministic multiplication processes induce the following two essential results:

(A) A dramatic increase in positive skewness — an essential criterion for Benford behavior.

(B) An increase in the order of magnitude — an essential criterion for Benford behavior.

Computer simulations of multiplications of five Normal distributions, each denoted as Normal(mean, standard deviation), yielded a highly skewed quantitative configuration as well as near-Benford digital results. A total of 35,000 realizations were obtained. The multiplication process is Normal$(750, 270) \times$ Normal$(700, 222) \times$ Normal$(200, 70) \times$ Normal$(876, 350) \times$ Normal $(972, 300)$. First digits are $\{31.9, 14.5, 10.5, 8.7, 8.3, 7.5, 6.9, 6.2, 5.5\}$. Their SSD has a low value of 21.6.

Computer simulations of multiplications of four Uniform distributions using 35,000 realizations yielded a highly skewed quantitative configuration as well as near-Benford digital results.

The Uniform distribution is denoted as Uniform$(a = $ minimum, $b = $ maximum). The multiplication process is Uniform$(3, 40) \times$ Uniform$(2, 33) \times$ Uniform$(7, 41) \times$ Uniform$(1, 29)$.

First digits are {30.3, 18.0, 12.9, 9.5, 7.7, 6.5, 5.6, 4.9, 4.6}. Their SSD has a very low value of 0.6.

Hence, even though the processes use highly symmetrical and anti-Benford distributions, the resultant quantitative configurations are highly skewed, and their digital results are nearly Benford.

Chapter 13

Addition Processes Lead to the Symmetrical Normal and away from Benford

Let us demonstrate the sharp contrast between addition and multiplication in terms of the resultant quantitative configuration and the resultant OOM. This shall be accomplished by converting the 10-by-10 multiplication table into an addition table.

We may view the table as a tool for those who feel too tired to do additions quickly in their heads, just as the multiplication table is memorized in youth as a tool to be used later in life.

The entire range of [2, 20] is partitioned into six equitable quantitative sections of three-unit width each: $\{2,3,4\}$, $\{5,6,7\}$, $\{8,9,10\}$, $\{11,12,13\}$, $\{14,15,16\}$, $\{17,18,19\}$, with the value of 20 conveniently excluded and where the sections are initially empty. Then, we let the values from the addition table fall carefully within each section according to their sizes. The goal is to group the 99 sums of the addition table according to sizes.

Figure 12 demonstrates such quantitative partitioning of the entire additive territory in detail. Figure 13 depicts the histogram of the number of values falling within each section.

The histogram is not skewed but rather fairly symmetric, and quantitative proportions are concentrated mostly around the middle part of the range. This is not a coincidence, but rather it signifies a very persistent trend found in all addition processes. The central limit theorem (CLT) points to the development of a symmetrical Normal

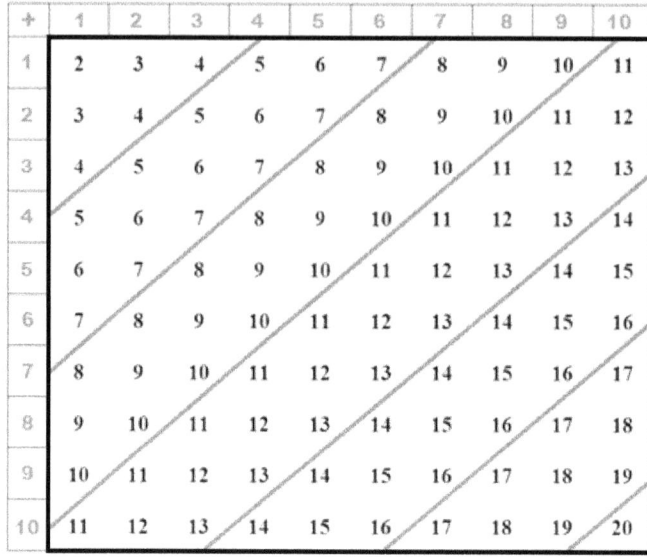

Figure 12. Quantitative Territorial Partitioning of the Addition Table

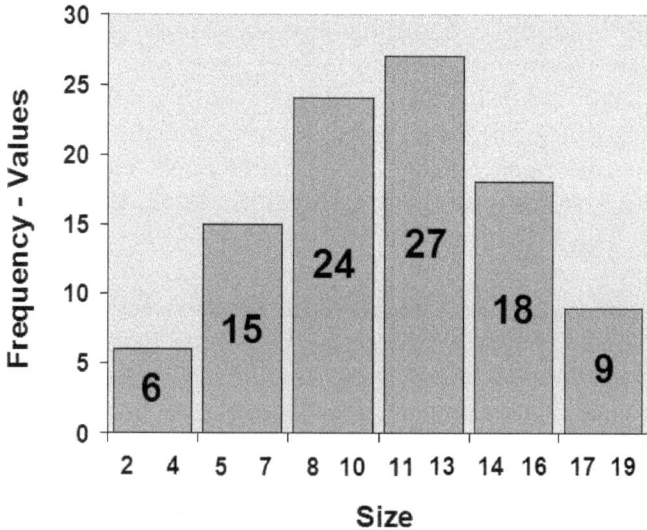

Figure 13. Histogram of the Addition Table, Showing Normal-Like Symmetry

distribution as the eventual distribution after numerous additions of random variables.

Crucially, the OOM for this addition table does not increase. Its maximum value is 20, and its minimum value is 2, hence the resultant post-addition OOM value of the entire table is $\text{Log}(20/2) = \text{Log}(10) = 1$, which is exactly the same as the OOM value of the original variable being added, namely the discrete set $\{1, 2, 3, 4, 5, 6, 7, 8, 9, 10\}$, with $\text{Log}(10/1) = \text{Log}(10) = 1$.

Hence, when the focus is on quantitative configurations, it can be said in extreme generality that addition processes favor the medium over the small and the big. When the focus is on first-digit configurations, nothing in general can be said *a priori* about addition processes, and their digital configurations depend on the specifics of the added variables and especially on the defined ranges of the added variables. It should be noted that additions are actively detrimental to the Benford configuration, in the sense that adding several highly Benford variables or data sets leads to a resultant additive process which is digitally not Benford and is quantitatively Normal-like. Incremental random additions of a Benford data set, DS, with itself over and over again in stages, as in DS, DS + DS, DS + DS + DS, and so forth, steadily lead away from Benford and skewness and toward the Normal configuration.

In general, random and deterministic addition processes do not induce any results that are essential to the criterion for Benford behavior:

(A) Lacking any increase in skewness, and moreover, on the contrary, actively increasing the symmetry of the resultant distribution, with added concentration forming around the center.

(B) Lacking any increase in order of magnitude beyond the existing maximum order of magnitude within the set of added variables.

Chapter 14

The Multiplicative Central Limit Theorem and Lognormal Distribution

In mathematical statistics, the sum of numerous independent realizations from an identical random variable X is known to be the Normal distribution, in the limit, as the number of these added realizations becomes large enough and almost regardless of the distribution form or parameters of X. This seminal result, which is called the central limit theorem (CLT), can also be applied — with some limitations — to the sum of different and distinct random variables, and moreover — under particular conditions — even for dependent realizations, and all this significantly enlarges its scope. If X_J represents one particular realization from the random variable X, then Normal $= X_1 + X_2 + X_3 + \cdots + X_N$ in the limit as N increases. In addition, the Lognormal distribution is defined as a variable whose natural logarithm is the Normal distribution, namely:

Lognormal(location, shape) $= e^{\text{Normal}(m=\text{location, sd}=\text{shape})}$

The shape and location parameters of the Lognormal are the standard deviation (sd) and the mean, respectively, of the generating Normal distribution. This definition, together with the CLT, implies that the Lognormal distribution can be represented as a process of repeated multiplications of a random variable, namely as

Lognormal $= e^{[\text{Normal}]} = e^{[X_1 + X_2 + X_3 + \cdots + X_N]}$
Lognormal $= (e^{X_1}) \times (e^{X_2}) \times (e^{X_3}) \times \cdots \times (e^{X_N})$

This result demonstrates that repeated multiplications of a random variable (of the form e to the power of any random variable)

is Lognormal in the limit as N becomes large, and this is called the multiplicative central limit theorem (MCLT).

The Normal distribution is obtained in repeated additions of a random variable.

The Lognormal distribution is obtained in repeated multiplications of a random variable.

Yet, not all multiplication processes and Lognormals are created equal; there is one decisive exception where multiplications actually behave just like additions! This surprising case occurs whenever the multiplicative process is composed exclusively of random variables with an exceedingly low OOM and where results are neither digitally Benford nor quantitatively skewed but rather quite symmetrical, resembling the Normal distribution. As a demonstration of this surprising counterexample, we examine the 70-to-79 multiplication table (a variation of our standard 10-by-10 multiplication table) using the particular set of 10 integers {70, 71, 72, 73, 74, 75, 76, 77, 78, 79} to be multiplied by themselves, with the smallest value of $70 \times 70 = 4900$ and the biggest value of $79 \times 79 = 6241$. Surprisingly, this table is not skewed at all but rather nearly symmetrical and Normal-like, while the first digits do not even remotely resemble the Benford configuration but are rather exclusively of {4, 5, 6}. The reason why this multiplicative process failed to achieve Benfordness and skewness is because, here, $\text{OOM} = \text{Log}(79/70) = \text{Log}(1.13) = 0.053$, which is very low, and also because the process is composed of only two such weak multiplicands, as opposed to numerous weak ones, which together might have perhaps helped in leading to a higher OOM. The resultant OOM of this multiplication table is still quite low, namely $\text{Log}(6241/4900) = 0.105$. Nonetheless, in general, the MCLT could still be applied if there are many variables of a low order of magnitude involved (i.e., N is sufficiently large for convergence), and it still guarantees that the resultant distribution is Lognormal, albeit with a low shape parameter, appearing symmetrical and Normal-like, and where the digits are not Benford. This possibility renders multiplication processes of low-variability data sets and variables non-Benford. If one wishes to overcome the exceedingly low OOM of the multiplicative variable, then an exceedingly high number of multiplications needs to be performed (i.e., N must be enormous, much more so than the usual number needed for convergence to Lognormal), which at long last could result in Benfordness and skewness and a sufficiently high resultant OOM. It should also be noted carefully that the opposite pole or reverse extreme here is the case of multiplications of only two or three symmetrical random variables, such as Normal or Uniform, each with an extremely high OOM, where the MCLT does not apply for a lack of a sufficient number of multiplications (i.e., N is too small for convergence) and which still yields the Benford digital configuration, quantitative skewness, and a very high resultant OOM, yet the resultant data are not Lognormal. These odd and extreme cases, such as the above, and other facts teach us that, in extreme generality:

(1) The MCLT (i.e., Lognormal) does not guarantee Benford behavior in multiplication processes, since variables with an exceedingly low OOM might preclude Benford behavior.

(2) The MCLT (i.e., Lognormal) is not necessary for Benford behavior in multiplication processes, since variables with an exceedingly high OOM lead rapidly to Benfordness with just two, three, or very few multiplications.

(3) The resultant OOM of the entire multiplicative process is the crucial and only factor in determining Benford behavior and skewness in multiplication processes.

(4) The resultant OOM of the entire multiplicative process springs from the confluence of two factors: the number of multiplications performed, and the individual OOMs of the variables themselves.

(5) The resultant OOM of the entire multiplicative process is simply the sum of the OOMs of the individual variables (either identical or distinct) being brought forward to be multiplied at each step of the multiplicative process. This sum must be over 3 in order to obtain sufficient positive skewness and Benford behavior for the entire multiplicative process.

Note: Simulations of the Lognormal distribution for the purpose of examining its digital and quantitative behavior reveal that a near-perfect Benford behavior and highly skewed quantitative configuration are obtained whenever the shape parameter is approximately over 1.2, and this result is totally independent of the value assigned to the location parameter. On the other hand, the Lognormal is decisively non-Benford, symmetrical, and Normal-like, for shape values less than approximately 0.4. This is consistent and expected since it is exactly the shape parameter (shp) that determines the OOM value of the Lognormal, not the location parameter (loc), as can be seen from the very definition of the Lognormal, where the standard deviation (shape) of the generating Normal determines the variability and spread of that Normal, and the generating Normal here exists only in order to produce log values (exponents) for the Lognormal. Hence, a high shape parameter (standard deviation) implies high variability in these exponents of the Lognormal. The decimal logarithm of the Lognormal is Normal(loc $\times$ $Log_{10}(e)$, shp $\times$ $Log_{10}(e)$), and the three-sigma rule implies that values fall from $[(loc) \times Log_{10}(e) - 3 \times (shp) \times Log_{10}(e)]$ to $[(loc) \times Log_{10}(e) + 3 \times (shp) \times Log_{10}(e)]$. So, the log-axis span is $6 \times (shp) \times Log_{10}(e)$, which constitutes the value of OOM and is ideally over 3, hence $6 \times (shp) \times Log_{10}(e) > 3$, and therefore shp > 1.15.

Chapter 15

Multiplications are More Prevalent than Additions in Real-Life Data

The scope of applications and favorable outcomes in terms of encountering the Benford phenomenon in real-life physical data due to the involvement of some multiplicative processes (and the avoidance of additive influences) is truly enormous, encompassing all scientific disciplines. The customary multiplicative form of the vast majority of the equations and laws in physics, chemistry, astronomy, economics, biology, engineering, and other disciplines, as well as the multiplicative form of the numerous expressions relating to their applications and results, all lead to the manifestation of quantitative skewness and often also to the Benford phenomenon in the natural sciences and in real-life data. Since these data sets are often with a high enough OOM and variability, they are frequently Benford.

Isaac Newton gave us $F = M \times A$, not $F = M + A$. Hence, the derivations and results due to the most famous expression in physics (Force) = (Mass) × (Acceleration) are related to multiplicative processes, not to additive processes. Newton also gave us the law of universal gravitation, $F_G = G \times M_1 \times M_2 / R^2$, which is written in multiplicative and divisional forms and not in additive or subtractive forms, such as, say, $F_G = G + M_1 + M_2 - R^2$.

The following example is one manifestation from classical mechanics, which points to multiplicative processes resulting in a Benford configuration. This example helps in demonstrating that such situations occur quite frequently in scientific and physical data.

The example considers the final position of numerous particles of random mass M, thrown linearly from the same location at random initial speed V_I, under a decelerating random force F, until each one comes to rest. In other words, each particle comes with a random and unique combination of M, F, and V_I. The equations in physics describing the motion are:

(i) $V_{\text{FINAL}} = V_I + A \times T$, which leads to $T_{\text{REST}} = V_I/A$ as the time it takes to achieve rest;

(ii) $F = M \times A$;

(iii) Displacement $= V_I \times T - (1/2) \times (A) \times (T^2)$. Hence, Displacement $= V_I \times (V_I/A) - (1/2) \times (A) \times (V_I/A)^2$, that is, Displacement $= (V_I)^2/(2 \times (F/M))$.

Monte Carlo simulations show strong Benford behavior here whenever M, F, and V_I are randomly chosen from Uniform$(0, b)$ for any value b. We intentionally choose the parameter a as 0 to ensure a high OOM.

Schematically, this entire arrangement is of the following form: Simulation $= (\mathbf{U_1} \times \mathbf{U_1})/(\mathbf{2} \times (\mathbf{U_2}/\mathbf{U_3}))$, and so we cannot apply the MCLT, and the distribution is not Lognormal, due to the very low number of multiplicands; nevertheless, due to the high OOM of the multiplied variables, the digits here are Benford.

Chapter 16

Tugs of War between Addition and Multiplication

The resultant POM (physical order of magnitude) of **multiplications** of several *identical or distinct* random variables is simply the product of the POMs of each of the participating variables. For example, for the multiplicative process $X \times X \times X \times Y$, where the variable X has a POM of 2, and the variable Y has a POM of 5, the resultant POM of the entire random product of the four variables is simply $\text{POM}_{\text{PRODUCT}} = 2 \times 2 \times 2 \times 5 = 40$. By definition, POM > 1, since the maximum is bigger than the minimum in the ratio Maximum/Minimum, therefore, resultant POM always increases with each extra participating multiplicand (variable). In concise mathematical notation, the resultant POM of the product of several J participating random variables is

$$\text{POM}_{\text{PRODUCT}} = \prod \text{POM}_J$$

Taking the decimal logarithm of both sides of the equation and applying the logarithmic identity $\text{Log}(A \times B) = \text{Log}(A) + \text{Log}(B)$ lead to the verbal rule that the OOM of the entire multiplicative process is simply the sum of the OOMs of the individual variables being multiplied.

The resultant POM of **additions** of several *identical* random variables is simply the POM of any one of the individual variables, with no increase, i.e.,

$$\text{POM}_{\text{SUM}} = \text{POM}_{\text{THE IDENTICAL VARIABLE}}$$

The resultant POM of **additions** of several *distinct* random variables is constrained to be less than or equal to the maximum (or the biggest) POM value in the set of all these distinct variables, so that there is no increase from that maximum level in the process of additions, i.e.,

$$\text{POM}_{\text{SUM}} \leq \text{POM}_{\text{MAX}}$$

Taking the decimal logarithm of both sides of the inequality leads to the verbal rule that the OOM of the entire additive process is not greater than the largest OOM within the set of individual variables being added.

At times, in real-life data, multiplication and addition processes mix together within one measurement or within the same mathematical expression, and consequently they compete fiercely for dominance, each attempting to exert the greatest influence upon sizes, quantities, and digits. While addition favors the medium and dislikes Benford, multiplication prefers the small and is often Benford. Neither is willing to compromise.

As an example relevant to real-life accounting and financial data, a single bill for one typical shopper in a large supermarket or at a big retail store may read as follows:

$3 \times (\$2.75 \text{ bread}) + 5 \times (\$2.50 \text{ tuna}) + 2 \times (\$7.99 \text{ cheese}) = (\$36.73 \text{ total bill})$

As another example relevant to real-life scientific and chemical data, the weight of a complex chemical molecule is derived from the **linear combination** of its constituent atoms.

For example, lactose $(C_{12}H_{22}O_{11})$ has a molar mass of 342.29648 g/mol. This particular molecular weight is derived from the following linear combination:

$12 \times (\text{Carbon Mass}) + 22 \times (\text{Hydrogen Mass}) + 11 \times (\text{Oxygen Mass})$

$= (\text{Lactose Mass})$

$12 \times (12.01070) + 22 \times (1.00794) + 11 \times (15.99940) = 342.29648 \text{ g/mol}$

Here, one multiplicand is the atomic weight in the periodic table, and the other multiplicand is the number of repeated atoms per distinct element within the complex molecule.

These two benign-looking arithmetical operations, namely addition and multiplication, in spite of their noble appearances, do occasionally go to war with each other over digital and quantitative configurations. The following describes six distinct random processes, where U represents Uniform(5, 33), and where all Us are distinct realizations from this uniform, so that for example, U × U + U × U utilizes four distinct realizations from this Uniform:

U + U
U × U + U × U
U × U × U + U × U × U
U × U × U × U + U × U × U × U
U × U × U × U × U + U × U × U × U × U
U × U × U × U × U × U + U × U × U × U × U × U

Here, the number of multiplicands in the process is increasing while the number of addends is fixed at 2. Six Monte Carlo computer simulations — with 35,000 runs each — are performed separately for each of the six processes. It should be noted that for the expression Uniform(5, 33)*Uniform(5, 33), say, two separate and independent realizations are needed from Uniform(5, 33), followed by the multiplication of these two realizations, and that we are not referring to the simple squaring of a single realization, such as in Uniform(5, 33)^2.

The simulation results are as follows:

U + U	$\{6.3, 19.2, 31.5, 26.8, 13.9, 2.3, 0.0, 0.0, 0.0\}$	SSD = 1360.0
UU + UU	$\{21.1, 5.7, 9.4, 11.4, 12.4, 11.8, 10.5, 9.4, 8.2\}$	SSD = 334.4
UUU + UUU	$\{42.4, 17.1, 6.7, 4.6, 5.2, 5.6, 6.3, 6.1, 5.9\}$	SSD = 222.5
UUUU + UUUU	$\{30.8, 22.4, 14.7, 9.5, 6.7, 5.1, 4.1, 3.5, 3.3\}$	SSD = 38.9
UUUUU + UUUUU	$\{26.0, 17.7, 13.9, 11.2, 8.9, 7.2, 6.0, 5.1, 4.1\}$	SSD = 22.3
UUUUUU + UUUUUU	$\{30.2, 16.2, 11.5, 9.8, 8.3, 7.1, 6.4, 5.6, 5.0\}$	SSD = 4.3

Finding addition sleeping at the wheel, being fixed at 2, multiplication then strives hard to win by gradually increasing the number of multiplicands from 1 to 6, and finally it is able to achieve the near-Benford condition within the last expression, where multiplication manifests itself 6 times versus only 2 manifestations of addition. Feeling overconfident now and conceited, multiplication then challenges addition to a new tug of war, willing to face 3 addends, given that it is already in possession of 6 multiplicands:

UUUUUU + UUUUUU + UUUUUU
$\{35.8, 16.4, 9.6, 7.3, 6.5, 6.6, 6.5, 5.8, 5.5\}$ SSD = 51.9

Upon seeing the severe setback in digital configuration, where the SSD is now over 50, multiplication deeply regrets its previous offer, and it then asks addition for permission to use two more multiplicands, achieving a total of 8 multiplicands versus these 3 addends:

UUUUUUUU + UUUUUUUU + UUUUUUUU
$\{27.2, 18.5, 13.7, 10.4, 8.5, 6.7, 5.6, 4.8, 4.6\}$ SSD = 12.1

Multiplication is now quite satisfied with this latest improvement in Benfordness, while addition is not much disturbed upon seeing that digits are getting back fairly close to Benford. Addition believes that ultimately its favorite Normal distribution will be encountered and that it will win over multiplication in due time. Nonetheless, addition wishes to make one last stand in order to flaunt its clout and demonstrate its ability to control the situation, and it expands its addends by just one more, namely a total of 4 addends versus these 8 multiplicands:

UUUUUUUU + UUUUUUUU + UUUUUUUU + UUUUUUUU
$\{25.3, 15.5, 13.7, 11.3, 9.5, 7.9, 6.3, 5.5, 4.8\}$ SSD = 35.8

Indeed, by adding one more addend to the process, addition was able to move it decisively away from Benford, managing to increase SSD from 12.1 to 35.8.

Where would all these actions and reactions, attacks and counterattacks, and the endless tit-for-tat war of attrition between addition and multiplication lead to? In extreme generality, the CLT is on the side of addition, leading typically to its eventual triumph. Surely, multiplication could win some battles in the short run, but it would normally lose the war in the long run as the resultant data finally become Normally distributed and the digital configuration turns non-Benford. Ultimately, the details of the structure and arrangement of the algebraic expression involved in such tugs of war determine the outcome, and the absence of a sufficient number of additive elements might actually preclude the full or even partial application of the CLT.

The **CLT's Achilles' heel** — in terms of its rate of convergence to the Normal — is the adverse possibility that added variables are highly skewed and come with a very high OOM, constituting a quite bad combination for the CLT and a challenge that needs to be overcome. Except for Uniforms, Normals, and other symmetrical distributions, which converge to the Normal quite rapidly after several or a few additions regardless of the value of the OOM of the added variables, all other asymmetrical (skewed) distributions show a distinct rate of convergence depending on the value of their OOM. For skewed variables with a very high OOM, the CLT can manifest itself with difficulties and very slowly, only after a truly large number of additions of the random variables. On the other hand, when skewed variables are of a low OOM, the CLT achieves near Normality quite rapidly after only a few or several additions. Fortunately, for revenue and expense accounting data as well as for molar mass data on molecules in chemistry, the CLT's Achilles' heel saves them from deviation from Benford in spite of the fact that some addition terms are involved, and this is so because price lists and catalogs as well as the periodic table are both fairly skewed, and both come with a relatively high OOM, which induces resistance to the CLT and Normality, resulting in Benford digital configuration and skewness.

The moral of the Achilles' heel story is that "not all perfectly Benford data sets are created equal!" Those with a very large OOM are by far more resistant to detrimental additive processes in the system, while those with just a sufficient OOM for (nearly) perfect Benford behavior are more vulnerable to random additions. A data set with, say, 500,000 points generated via simulations from the Lognormal with a shape parameter of 3.5 and whatever location is by far superior and much more resistant to random additions and the CLT than a comparable set with a shape parameter of, say, 1.2. Both data sets measure equally for all practical purposes in their present Benford status, yet the former is much more resistant and stable when compared to the latter.

Let us summarize in extreme generality the effects of random arithmetical processes on resultant data, excluding the considerations regarding their OOMs, the exact numbers of arithmetical applications involved, and symmetry or skewness conditions, all of which might change the presumed outcomes outlined as follows:

Randomly multiplying Uniforms yield Lognormal
Randomly multiplying Normals yield Lognormal
Randomly multiplying Lognormals yield Lognormal
Randomly multiplying Benfords yield Benford
Randomly multiplying non-Benfords yield Benford

Randomly adding Uniforms yield Normal
Randomly adding Normals yield Normal
Randomly adding Lognormals yield Normal
Randomly adding non-Benfords yield non-Benford
Randomly adding Benfords yield non-Benford

Addition Processes
Same POM — More Symmetry — CLT — Normal — Anti Benford

Multiplication Processes
More POM — More Skewness — MCLT — Lognormal — Pro Benford

Chapter 17

Partitions Typically Lead to Positive Skewness and Often to Benford

The mathematical field of integer partition investigates the ways an integral quantity can be expressed as a sum of positive integers. It deals with very simple and straightforward questions such as, "In how many ways and exactly how can the quantity 5 be broken into integral parts?" The answer to this question is an exhaustive list of all possible integral partitions of 5, detailed as follows:

5
4 + 1
3 + 2
3 + 1 + 1
2 + 2 + 1
2 + 1 + 1 + 1
1 + 1 + 1 + 1 + 1

Figure 14 demonstrates all possible integer partitions of 5 and the organization of all the parts neatly according to size via a histogram of sorts. The small clearly outnumbers the big here, and surely this is true not only for 5 but also for any integer N chosen to be similarly partitioned, where histograms are always positively skewed with a tail falling on the right. Yet, the artificial and unnatural limitation of restricting partitions only along integral forms precludes Benford behavior.

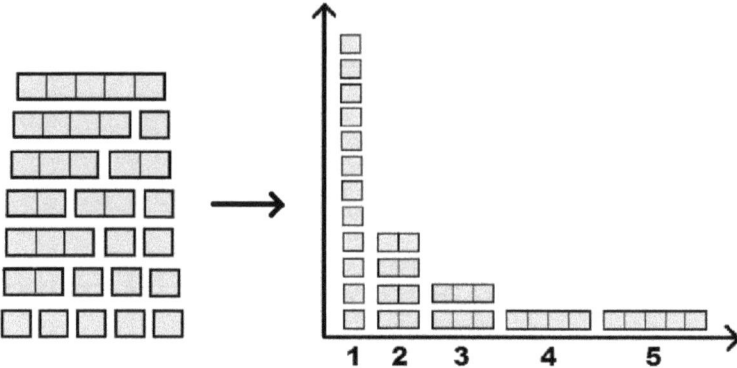

Figure 14.   All Possible Integer Partitions of 5 Leads to a Skewed Histogram

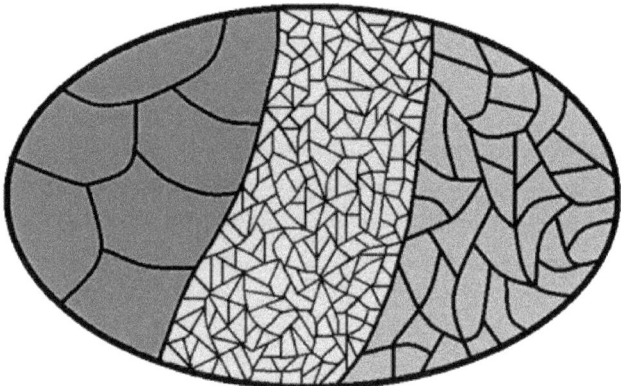

Figure 15.   An Equitable Mix of Small, Medium, and Big Leads to Skewness

Figure 15 provides another visual and intuitive demonstration that in a given natural partition containing the big, the small, and the medium, and where all three sizes exist equitably, many more small pieces should be found than big pieces. Although Figure 15 focuses on a two-dimensional area as the quantitative variable, the lesson learned from it is generic and applicable to any other type of quantity. The figure depicts one possible partition in the natural world where approximately one-third of the entire oval area consists of big parts (around the left-hand side), approximately one-third of the entire oval area consists of small parts (around the center), and approximately one-third of the entire oval area consists of medium parts (around the right-hand side), thus endowing equal portions of the overall quantity fairly to each size without any bias. The small decisively outnumbers the big here, and any crude histogram constructed out of Figure 15 would definitely fall on the right side. Surely, in nature, there exists no such order and grouping by size around the left–center–right sections or along any other dimensions. In nature, the big, the small, and the medium are all mixed in chaotically. But the order for the sizes along the left–center–right sections shown in this figure is made for pedagogical purposes: to visually reinforce for the reader the profound quantitative consequences affecting sizes in typical partition models.

In addition, real-life random, natural, and spontaneous partition processes never yield only particular sizes, never insisting that results should be of only three sizes. It's only for pedagogical purposes and to emphasize the generic lesson learned here that the sizes fell neatly into one of three categories and that all parts (mini areas) within each size category were made artificially of the same magnitude approximately, repeating themselves almost. In reality, chaotic and natural partitions yield a set of totally distinct sizes, without any categories, without repetitions, and with as many sizes as the number of parts. The artificial and unnatural limitation of having only three sizes, as in Figure 15, precludes Benford behavior, although positive skewness prevails.

These two examples above demonstrate a very profound, universal, and yet extremely simple principle regarding how a conserved quantity can be partitioned into parts, namely the generic observation that "**one big quantity is composed of numerous small quantities**", or, equivalently, "**numerous small quantities are needed to merge into one big quantity**".

Chapter 18

One-Dimensional Random Staged Partition

One well-structured arrangement of repeatedly partitioning a single quantity randomly into many parts is coined "random staged partition". This process is nicely exemplified by randomly breaking a big rock in multiple stages into much smaller pieces, applying the random continuous distribution Uniform(0, 1) as the percent determining each breakup proportion. The one-dimensional quantity being partitioned is not restricted to the weight or mass of a physical object, but rather this process refers to any generic one-dimensional quantity, such as weight (of a rock), length (of a pipe), time, angle, and energy, or any abstract mathematical quantity. Let us illustrate random staged partition with one concrete example of a three-stage process.

A **500** kilogram rock is broken in the first stage via the random 18%−82% percentage pair, yielding two new pieces: $[500] \times [18/100] =$ **90** and $[500] \times [82/100] =$ **410**. In the second stage, first, the 90 piece is broken via the random 71%−29% percentage pair, yielding two new pieces: $[90] \times [71/100] =$ **63.9** and $[90] \times [29/100] =$ **26.1**. Second, the 410 piece is broken via the random 86%−14% percentage pair, yielding two new pieces: $[410] \times [86/100] =$ **352.6** and $[410] \times [14/100] =$ **57.4**. In the third stage, each of the four pieces is again randomly broken. Figure 16 depicts the process from the original 500 kilogram rock to the end of the third stage. It should be noted that the original singular quantity is obviously always conserved after any

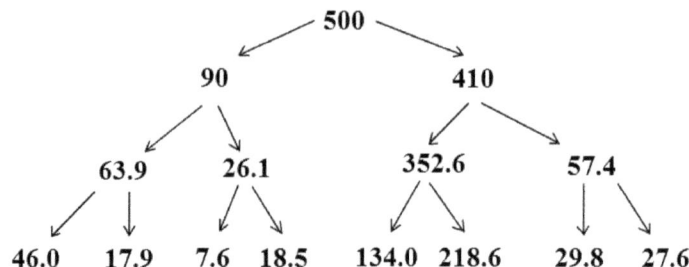

Figure 16. Random 500-kilogram Rock Breaking in Three Stages

number of stages, without any changes, so that in this example, the sum of all the pieces is always 500 kilograms at the end of any stage.

Random staged partitions always converge to Benford quite rapidly, and decisively so. After about 10 stages, having 2^{10} or 1,024 pieces, the SSD is almost always below 10. After about 14 stages, having 2^{14} or 16,384 pieces, the SSD is almost always below 1, and a near-perfect Benford configuration is achieved, while the quantitative configuration of the pieces is highly skewed, the small decisively outnumbers the big, and the histogram is falling to the right. Beyond 14 or 15 stages, saturation sets in, and no further reduction in the SSD is possible with additional stages.

It is beneficial to visualize how quantities here are evolving algebraically. This is accomplished by writing down the process carefully, stage by stage. Assuming the original weight of the rock as 1 kilogram, U_J as the Jth realization from the Uniform(0, 1) in the simulation scheme, and $(1 - U_J)$ as its complement, we have the following:

$\{1\}$

$\{U_1, \ (1 - U_1)\}$

$\{U_2 \times U_1, \ (1 - U_2) \times U_1, \ U_3 \times (1 - U_1), \ (1 - U_3) \times (1 - U_1)\}$

$\{U_4 \times (U_2 \times U_1), \ (1 - U_4) \times (U_2 \times U_1), \ U_5 \times (1 - U_2) \times U_1, \ (1 - U_5) \times (1 - U_2) \times U_1,$
$U_6 \times U_3 \times (1 - U_1), (1 - U_6) \times U_3 \times (1 - U_1), U_7 \times (1 - U_3) \times (1 - U_1), (1 - U_7) \times$
$(1 - U_3) \times (1 - U_1)\}$

And so forth to higher stages.

Clearly, it would be exceedingly rare to find identical values for the pieces here. For example, it is possible, in principle, to have $U_4 \times (U_2 \times U_1) = U_6 \times U_3 \times (1 - U_1)$, but this would require a very particular set of choices of U_J's, which is extremely unlikely and rare, having zero probability in a formal mathematical sense, since Uniform(0, 1) contains unaccountably infinite real numbers. The formal mathematical approach aside, and in a practical sense, for the partition practitioner or the data analyst assisting the process, it would be exceedingly rare to find two nearly identical pieces of almost the same size, such as $U_4 \times (U_2 \times U_1) \approx U_6 \times U_3 \times (1 - U_1)$, unless they are just a bit similar but definitely of distinct values. Certainly, there would exist no small–medium–big categories, or any other categories. There would be as many sizes as the number of pieces.

The above sequence of algebraic expressions demonstrates that a random staged partition can also be thought of as a multiplicative process, albeit with some dependencies between the terms, since, say, U_1 appears in each term and in all the stages, while U_2 or U_3 appear in each term from the third stage on, and so forth, mixing with the other U_J occurrences. As seen in earlier chapters, multiplication processes are known to be highly skewed quantitatively and have strong Benfordian digital tendencies. Thus, we expect (and get) skewness and Benford behavior here. This assertion relies on the facts that the arithmetical terms of the process involve numerous multiplicands, not just a few, and that multiplicands come with a sufficient OOM. Since we require at least 10 stages or more, we get 1,024 pieces after the 10th stage, and thus we guarantee having numerous multiplicands in the algebraic expressions.

The use of Uniform(0, 1), which supposedly possesses an infinitely large OOM, guarantees that the process can easily overcome any possible challenge (if any) from the algebraic dependencies between U_J occurrences. The OOM here is naively calculated as $\text{LOG(POM)} = \text{LOG}(1/0) = \text{LOG(Infinite)} = \text{Infinite}$. Yet, using the more robust and realistic CPOM measure yields $\text{LOG(CPOM)} = \text{LOG}((0.99)/(0.01)) = \text{LOG}(99) \approx 2$, which is not infinite yet sufficiently large. The random staged partition model cannot utilize any other random variables with a low OOM, such as Uniform(5, 7), because it needs to break a whole quantity into two parts via two fractional values, which together sum up to 1.00 and where each fractional value is between 0 and 1. Thus, this can only be achieved via Uniform$(0, 1)$, which is of sufficient OOM in this partition scheme, leading to Benford convergence.

Chapter 19

One-Dimensional Chaotic
Repeated Partition

Those who know Mother Nature well and are familiar with the way she works assert that she would probably never bother to carefully break her quantities in exact and deliberate stages and that random staged partition is not the typical process that she likes doing or is even capable of performing, since this process demands a great deal of mental effort and organization.

Let us imagine a real-life assembly line with workers and management, as in typical large corporations, attempting to physically perform the random staged partition process on a very long metal pipe. First, the workers decide on the initial random location where the pipe is to be cut, followed by the actual cutting. Then, designation tags of "1" and "2" for the two newly created pieces are made and placed on each piece. This is followed by the orderly cutting of piece "1" and then piece "2" at a random location within each piece, resulting in four pieces. Without such designation tags, there exists the possibility that, by mistake, one piece is cut twice, leaving the other piece intact. Next, the workers designate the four newly created pieces as "1", "2", "3", and "4". At this stage, piece designation becomes even more crucial to avoid confusion and mistakes, and to remember which pieces were already cut and which pieces are awaiting their turn. It is highly doubtful that Mother Nature is capable of or even interested in such a serious and rigid type of work.

How would temperamental Mother Nature go about breaking a rock or a pipe her way, leisurely, chaotically, and consistent with her free-spirit attitude and strong dislike of regimentation?

Her first act in the process is the breaking of the original rock into two pieces randomly via the continuous Uniform(0, 1) to decide on the proportions of the two fragments. There is no need whatsoever to designate any pieces with any tags thereafter, since the order of breakups does not matter to her in the least. Her second act is the totally relaxed and random selection of any one of the two pieces laying in front of her without much thought for which to choose, followed by its fragmentation into two pieces randomly via Uniform(0, 1) and the throwing of the two newly created pieces back into the pile carelessly, resulting in three pieces. Her third act is the totally relaxed and random selection of any one of the three pieces, followed by its fragmentation into two pieces randomly via the Uniform(0, 1) and the throwing of the two newly created pieces back into the pile carelessly, resulting in four pieces. If this continues for a sufficiently large number of steps, say, approximately over 15,000 steps, then the Benford configuration is decisively obtained, and nearly perfectly so, and where the pieces are quantitatively skewed as well. Thereafter, saturation sets in, and no further reduction in the SSD is possible with additional steps. This process is termed "chaotic repeated partition".

Three Monte Carlo computer simulations of chaotic repeated partition are run, all starting with an initial 100-kilogram rock and with the first having 500 steps, the second having 1,300 steps, and the third having 20,000 steps. The random breakup ratio of Uniform$(0, 1)$ is applied. These three schemes yield the following results:

- Chaotic repeated partition 500 steps:
  $\{32.1, 19.4, 11.6, 8.6, 8.0, 5.8, 6.0, 4.0, 4.6\}$, SSD $= 11.4$
- Chaotic repeated partition 1,300 steps:
  $\{29.0, 18.6, 12.8, 9.4, 9.8, 6.1, 5.6, 4.7, 4.1\}$, SSD $= 6.6$
- Chaotic repeated partition 20,000 steps:
  $\{30.0, 17.2, 12.4, 9.9, 8.0, 6.8, 6.0, 5.1, 4.8\}$, SSD $= 0.3$
- **Benford's Law of first digits:**
  $\{\mathbf{30.1, 17.6, 12.5, 9.7, 7.9, 6.7, 5.8, 5.1, 4.6}\}$, **SSD $= 0$**

Let us examine visually how quantities here are evolving algebraically, which is accomplished by writing down one possible random scenario step by step. Assuming the original weight of the rock as W kilogram, U_J as the Jth realization from Uniform$(0, 1)$ in the simulation scheme, and $(1 - U_J)$ as its complement, we have the following:

$\{W\}$ W is broken
$\{U_1 \times W, (1 - U_1) \times W\}$ then only $U_1 \times W$ is broken
$\{U_2 \times U_1 \times W, (1 - U_2) \times U_1 \times W, (1 - U_1) \times W\}$ then only $(1 - U_1) \times W$ is broken
$\{U_2 \times U_1 \times W, (1 - U_2) \times U_1 \times W, U_3 \times (1 - U_1) \times W, (1 - U_3) \times (1 - U_1) \times W\}$ then $U_2 \times U_1 \times W$
$\{U_4 \times U_2 \times U_1 \times W, (1 - U_4) \times U_2 \times U_1 \times W, (1 - U_2) \times U_1 \times W, U_3 \times (1 - U_1) \times W, (1 - U_3) \times (1 - U_1) \times W\}$

And so forth to higher steps, producing many more broken pieces.

Clearly, after a sufficient number of such steps, on average each term (i.e., each piece) contains numerous algebraic multiplicands; therefore, the resultant set of pieces can be thought of as emerging from a multiplicative process with a sufficiently high OOM relating to Uniform$(0, 1)$, which leads decisively to Benford behavior — in spite of the dependencies.

Chapter 20

One-Dimensional Random Real Partition

A random and spontaneous one-dimensional partitioning process lacking any dependent stages leads to quantitative skewness and to an approximate or near-Benford configuration, while resultant pieces attain the form of the Exponential distribution. The process is labeled "random real partition", and the term "real" refers to the real number line or x-axis, expressing the length dimension. This process is nicely exemplified by the cutting of a one-dimensional linear and long pipe at random and independent locations along its length. As an example, a 15-meter-long metal pipe is to be randomly partitioned into 30 parts. This is accomplished by obtaining 29 independent random points along the pipe — prior to the moment of the actual cutting — to serve as marks indicating where the pipe should be partitioned. These marks constitute the grand plan of the entire partition process, being generated via independent realizations from the continuous Uniform(0, 15), endowing equal chance to all locations along the pipe anywhere equally and uniformly. This is followed by the actual cutting of the long pipe into 30 parts. Figure 17 depicts

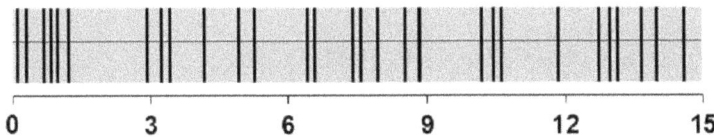

Figure 17. The 29 Random Marks along the 15-meter Pipe Prior to Partition

one such actual process with 29 marks along the pipe obtained via computer simulations.

A random real partition where the pipe or the generic quantity is thoroughly broken using many marks (far more than 29), thus having numerous resultant parts, is approximately Benford, but a limited partition with only a few marks, thus having a few resultant parts, is not in any way Benford. A partition with approximately over 5,000 or 10,000 parts is always approximately Benford, resulting in SSD of around the 8 to 10 level. Going further beyond that many parts never gets closer to Benford as saturation sets in. The set of lengths of these numerous pieces or parts resulting from a random real partition corresponds to the Exponential distribution, which is also known to be approximately Benford. For example, an L-meter pipe being randomly cut into N parts via $(N-1)$ random marks between 0 and L and utilizing the Uniform distribution implies a rate of $(N-1)/(L)$ marks per unit distance and points to an Exponential distribution with a lambda parameter value of $(N-1)/(L)$, assuming $N > 10,000$ approximately.

Another perspective on random real partition is its description as in the following scheme:

- Generate N realizations from the continuous Uniform$(0, P)$.
- Order them from low to high, add 0 on the very left, and add P on the very right.
- The data set is $\{0, U_1, U_2, U_3, \ldots, U_{N-2}, U_{N-1}, U_N, P\}$.
- Generate the differences data set out of the one above.
- This data set is $\{(U_1 - 0), (U_2 - U_1), (U_3 - U_2), \ldots, (U_{N-1} - U_{N-2}), (U_N - U_{N-1}), (P - U_N)\}$.
- This data set is approximately Benford as N approaches 10,000, and it is Exponentially distributed.

Performing a Monte Carlo empirical test on a random real partition and its compliance with Benford and applying the Uniform as the distribution generating the marks yields the following results:

Uniform$(0, 800)$ partitioned via 10,000 marks:
$\{32.2, 16.7, 11.3, 8.6, 7.5, 6.8, 6.0, 5.8, 5.1\}$.

The SSD value is **8.2**, and such a low value indicates that this partition is fairly close to Benford.

Here, there are 10000/800, or 12.5 marks per unit, and therefore this process corresponds to an Exponential distribution with a lambda parameter value of 12.5. In one computer simulation run for this Exponential distribution, first digits distribution came out as {32.3, 16.5, 11.4, 8.7, 7.5, 6.8, 6.3, 5.4, 5.1}.

Stronger convergence to Benford is obtained when the symmetrical Normal distribution is substituted for the continuous Uniform. An even stronger convergence is obtained with negatively skewed or positively skewed non-symmetrical distributions, Benford or otherwise. Surely, the mathematical correspondence to the Exponential distribution is ruined in all such non-Uniform substitutions of distribution. Four Monte Carlo simulation examples were obtained:

Normal(19, 4) partitioned via 35,000 marks:
{28.5, 17.4, 13.2, 10.5, 8.3, 6.8, 6.0, 5.0, 4.3}.

The SSD value is **4.0**, and this lower value (in comparison to the Uniform) indicates that applying the Normal in partitions is in general closer to Benford than the application of the Uniform.

$K \times X^3$ on (1, 50) partitioned via 25,000 marks:
{30.8, 17.5, 12.0, 9.1, 8.0, 6.8, 5.6, 5.4, 4.8}.

The SSD value is **1.3**, and this very low value indicates that partitioning along highly negatively skewed distributions leads to stronger Benford results in comparison to symmetrical ones.

k/x on (1, 10) partitioned via 30,000 marks:
{30.0, 18.0, 12.2, 9.4, 7.8, 6.9, 5.7, 5.2, 4.7}.

The SSD value is **0.4**, and such an exceedingly low value indicates that applying the positively skewed k/x distribution gets us extremely close to Benford.

Lognormal(9.3, 1.7) partitioned via 35,000 marks:
{30.0, 17.6, 12.4, 10.0, 8.0, 6.6, 6.0, 5.0, 4.4}.

The SSD value is **0.2**, and such an exceedingly low value indicates that applying the high-shape Lognormal gets us extremely close to Benford. Notation: $\text{Log}N$(location = 9.3, shape = 1.7).

In each of the above four Monte Carlo simulations, we first simulate thousands of realizations from the abstract distribution, each realization being a point or a location on the x-axis. We then sort them from low to high and calculate and record the distances between adjacent points (subtracting left from each right), followed by the examination of the first digits of these distances.

Benford is associated with randomness in a heuristic sense, hence the more randomness associated with a given process, the closer we get to the Benford configuration. The gradual labor-intense stages performed by the workers in random staged partition as well as the easy and carefree steps taken by Mother Nature in chaotic repeated partition, both led to nearly perfect Benford configurations, as each new random stage or step was built upon and dependent on all previous stages or steps — and which could perhaps be considered as building even more randomness upon existing randomness. In contrast, all marks of a random real partition could, in principle, come into being simultaneously in one epic instant, while these marks are definitely independent of each other and uniformly spread. So, all these could be considered a bit less random in a sense and perhaps the reason why the process does not achieve full Benfordness. Yet, another heuristic argument is needed to explain why the Normal, KxX^3, k/x, and the Lognormal, all led directly to Benford. One justification or rather rationalization could be put forward by arguing that their more complex and uneven spread along the x-axis constitutes an increase in randomness, as opposed to the "more predictable" and more even spread of the Uniform.

Chapter 21

Two-Dimensional Random Partition

The most primitive or simplistic two-dimensional random partition of an oval-shaped tablecloth, soft fabric, or cardboard, where the cutting is performed in the freest possible way, without any restrictions, leads to skewness and the Benford configuration for the set of areas of the resultant pieces. Figure 18 depicts an example of an original oval shape partitioned along random curved paths.

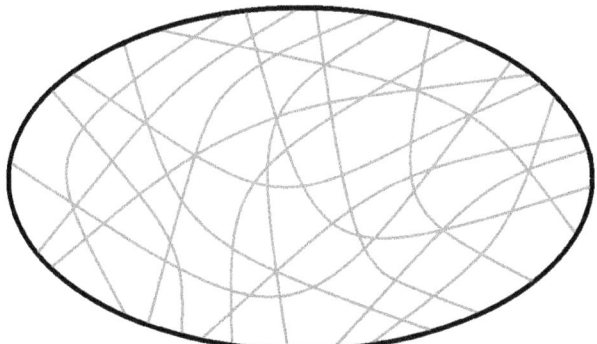

Figure 18.   Oval-Shaped Tablecloth is Partitioned along Curved Paths

A large and strong pair of scissors is imagined, being allowed to curve around while cutting without constraints. This is done by cutting from any edge to any other edge while turning around in any way one sees fit, without stopping in the middle. The overall oval shape stays intact after each such cutting act to let the next cutting act also be performed on the entire oval-shaped tablecloth, independently of all previous cuts, until the very end of the long cutting process. Then, when no more cutting is desired, when the scissor is put away for good, the still-standing fragile oval tablecloth consisting of its physically cut but touching parts is dismantled, after which the oval shape is no longer intact, and all the pieces are gathered into one big chaotic pile, followed by the recording of the areas of the pieces. Such restriction-less, natural, and spontaneous partition always leads to the Benford configuration and quantitative skewness, assuming numerous cuts have been applied, yielding many resultant pieces.

The simple-minded and uneducated person tasked with cutting the tablecloth may not have mastered the entire 10-by-10 multiplication table at school, yet, in reality, that person is subconsciously performing several highly sophisticated multiplications all at once. As the scissors advance from one edge to another, each area A they encounter along the way is being split into $(A) \times \text{Uniform}(0, 1)$ and $(A) \times (1 - \text{Uniform}(0, 1))$. Hence, each full cut from edge to edge represents in essence random multiplicative processes for several areas at once. Yet, in reality, our naïve and unsophisticated worker is just cutting, not multiplying!

One controversial partition interpretation in Benford's Law is the attempt to model the data of the time between all successive earthquakes worldwide for any given year, as a random partitioning of time. This could perhaps apply to the data set of 19,452 earthquakes that occurred during the year 2012, presented in Chapter 5. This view assumes the random partitioning of the entire time of one year into numerous time segments of peace and quiet when no earthquakes occur. Following another line of reasoning, namely the multiplicative view of partitioning, it then might be erroneously extrapolated and applied to the earthquake data, to be interpreted as a multiplicative process. Yet surely, neither the description of partitioning nor the description of multiplications is what earthquakes are all about. The more specific, daring, and fairly radical interpretation of the entire data on 2012 global earthquakes as a random real partition might be suggested or attempted via the application of Uniform(0, 365) in units of days, as time since midnight on January 1st, obtaining 19,452 random independent realizations representing occurrences of earthquakes, then sorting values from low to high and subtracting adjacent values, letting differences be thought of as the time intervals between earthquakes. Surely such a description or interpretation is not what earthquakes are all about, and in particular, the assumption of independent earthquake occurrences is highly problematic. Indeed, actual global earthquake data as well as geology theory demonstrate that the timings of earthquakes in close local proximity to each other and occurring during the same day or week are dependent on each other as they frequently correlate, and more specifically so for aftershocks, namely the sequence of earthquakes that occur after a larger mainshock on a fault.

Another controversial partition interpretation in Benford's Law is the attempt to model the data of population centers in the USA, considering, say, 19,000 cities and towns containing a total of, say, 300,000,000 people, as a random partitioning of the entire population, who are imagined to be initially and temporarily housed in a huge camp awaiting deportation, to be later placed into cities and towns in one epochal moment. The more specific, daring, and fairly radical interpretation of the population data as a random real partition might be suggested or attempted via the application of 18,999 independent realizations from Uniform(0, 300,000,000), rounding up fractional parts into integral values (i.e., people), then sorting values from low to high and subtracting adjacent values, letting differences be thought of as populations of cities and towns. Surely, such a description or interpretation is not what population centers are all about, and in particular, the assumption of independent population values is highly problematic, especially due to immigration and migration between cities. Each city has its own unique, random, and long trajectory of growth and history, as opposed to some simple realization from the Exponential.

Indeed, empirical examinations of actual population data as well as actual earthquake data reveal excellent agreement with the Benford configuration, with exceedingly low SSD values of less than 2 typically. This is much closer to Benford when compared to random real partitions and the Exponential, which yield higher SSD values of around 8 — being just close to Benford. Only desperate attempts to save the partitioning view by insinuating random staged partition or chaotic repeated partition as models of populations and earthquakes could bring empirical and theoretical SSD values into agreement, but such imaginary models with stages and steps are farfetched and definitely not suitable for the data.

Chapter 22

The General Requirements for Partitions to Converge to Benford

In extreme generality, the Benford digital configuration is found in quantitative partition models whenever: (I) partitions are performed in a purely random fashion, freely without any arbitrary restrictions; and (II) partitions are done repeatedly enough so that the original quantity is finally broken into numerous resultant parts. Adherence to the first of these two necessary rules, namely that partitioning is done in a truly random and free fashion, almost always guarantees two other consequences that are essential for achieving the Benford configuration, namely: breaking the original quantity along the real number basis and not exclusively along the integers, so fractional values are allowed; as well as ensuring that the original quantity is broken into totally distinct resultant parts of unique sizes, with almost no repetitions of sizes, and thus without the possibility of successfully grouping the resultant parts into some compelling and obvious size categories, as was done successfully in Figure 15, which is not Benford. A partition leading to Benford must thoroughly break the original quantity into all sorts of random, chaotic, and disparate sizes.

The surprising connection between quantitative partition models and Benford's Law was first discovered by the physicist Don S. Lemons in 1986 with the publication of his two-page seminal article regarding one particular partition model and the typical occurrences of positive skewness in data. This was later followed by more literature on quantitative partitions in the context of Benford's Law by the mathematician Steven J. Miller and this author.

Chapter 23

Benford Model for Planet and Star Formations

Intuitively, it is expected that the Benford configuration should also be found in construction (pieces of mass forming) just as it was found in destruction (partitions). This is so in spite of the arithmetic fact that constructions are typically associated with additions, while destructions are typically associated with multiplications. One such model of construction is of numerous pieces of mass being generated and thrown into space, each piece being generated with some random quantity derived from the anti-Benford type of Uniform(0, maximum weight), then thrown into space or a nebula, having a P probability of getting attached and glued to any other existing pieces there and a $(1 - P)$ probability of continuing to float out there in the nebula alone without connecting to any existing pieces. If the probability is indeed favorable for a newly created piece R with weight $W_{\mathbf{R}}$ and it is about to be attached to another piece, then it chooses any piece randomly with equal uniform discrete probability, regardless of the mass or weight of the existing pieces, and if that chosen piece Q has weight $W_{\mathbf{Q}}$, then the two fused pieces are now one bigger new piece of weight $(W_{\mathbf{R}} + W_{\mathbf{Q}})$, which is obtained via an additive operation. To recapitulate, a new piece of random weight is created and thrown into space containing older pieces, and then randomly deciding whether to merge with another existing piece chosen totally randomly or to continue its lone existence, followed by yet another creation of a new piece, and so forth.

The model assumes that even though pieces do interact via gravitational forces of attraction, which depends on the value of their masses (as per Newton's term $M_1 \times M_2$ in the numerator), effectively, with the absence of the disparity of extremely massive masses versus extremely light ones, that interaction is not so much a function of the mass value of the older pieces as predicated by the gravitational formula, but rather it is mostly a function of the random trajectory of an arriving piece, which may be aiming (P) or may not be aiming $(1 - P)$ at some particular existing piece. The gravitational force is inversely proportional to the distance squared (as per Newton's term $1/R^2$ in the denominator), thus the gravitational sphere of influence is really mostly in the immediate vicinity of the object. It follows that what matters most here is the trajectory of the incoming piece and how close it approaches any existing piece. In conclusion, any arriving piece is equally likely to choose from the existing pieces, almost regardless of weight, without preferring heavier ones, as its choice depends almost solely on its own trajectory.

Indeed, computer simulations show that such a process converges fairly close to Benford, assuming P is bigger than 0.50 approximately. Larger P values, such as 0.80 or 0.90, yield better results. The first piece of mass is just being put out there without the possibility of merging it with anything. It is only from the second piece of mass onward that the process starts in earnest. The condition $P > 0.50$ is assumed, and Benford configuration is obtained thus most pieces are fused together, and the original number of created and thrown pieces is much reduced in the process; this reduction is symbolically indicated as [(original # of created pieces) → (final # of pieces)] in Figure 19, which depicts the results for five distinct simulations having a variety of P values and initial number of pieces.

P = 0.97	[90000 →	2682]	31.5	18.0	13.2	9.1	7.0	6.1	6.1	4.7	4.4
P = 0.93	[40000 →	2669]	30.6	16.7	11.5	9.0	7.4	6.4	7.2	6.0	5.1
P = 0.90	[40000 →	3904]	29.6	15.0	13.1	10.6	7.3	7.3	5.6	6.7	4.9
P = 0.75	[15000 →	3882]	30.0	18.8	13.1	8.0	7.0	5.9	5.7	5.8	5.6
P = 0.67	[50000 →	16531]	31.3	17.7	11.2	7.7	6.7	6.2	6.3	6.3	6.6
Benford's Law 1st Digits:			30.1	17.6	12.5	9.7	7.9	6.7	5.8	5.1	4.6

Figure 19.   Digital Results of Simulations of Star and Planet Formations

If probability P is made to be too low, such as 0.2, then most of the pieces created fly out without getting attached to anything; almost no star or planet is being formed, almost no reduction in the number of pieces is observed, and the digit distribution of the final system is almost identical to that of the generating process, namely of the original Uniform configuration. If probability P is 1.00, then all the pieces necessarily get glued together and merge into one huge chunk of matter; the data set is of the tiniest size of 1, and only one unique first digit prevails.

For probability $P > 0.50$, the digit distribution converges to the Benford configuration after approximately 1,000 or 2,000 final pieces come into existence, then saturation sets in, and subsequent thrown pieces are not a factor any more as the Benford configuration that emerged persists, regardless of new arrivals. It should be noted that if we employ Benfordian-like distributions instead of the Uniform to generate the quantities of the pieces, such as the Exponential or the Lognormal with high shape, then the results are stronger, and the Benford convergence occurs earlier, and this is certainly expected and obvious. In conclusion, surprisingly, we have arrived at the Benford configuration in spite of applying a (unique and complex) addition process and in spite of generating quantities via the anti-Benford Uniform distribution! Our initial intuition that random construction as the complement twin or mirror image of random destruction should also yield the Benford configuration was correct!

Chapter 24

Consolidation and Fragmentation Processes

A consolidation and fragmentation process is a scheme where an initial large set of identical quantities (say, L balls all initially with the same identical weight W) constantly alternates between (I) random consolidation, namely the fusing of two randomly chosen balls involving the additive term $(\text{Ball}_1 + \text{Ball}_2)$, and (II) random fragmentation, namely the splitting of one randomly chosen ball involving the two multiplicative terms of $(\text{Ball}) \times \text{Uniform}(0,1)$ and $(\text{Ball}) \times (1 - \text{Uniform}(0,1))$. Each cycle consists of one consolidation and one fragmentation, randomly choosing balls to be consolidated and to be split, respectively, followed by many more such cycles, over and over again, so that some balls are getting heavier and bigger while others are broken into lighter and smaller balls, randomly.

The number of quantities (i.e., the number of balls L) in the system does not change, and it stays constant throughout the cycles, as it goes down by one and then up by one in each cycle. Also, the grand total of all the quantities in the system, namely the sum of the weights of all the balls, stays constant throughout the cycles without change, so that $\sum \text{Ball}_J$ is fixed at $L \times W$, as it was originally. In other words, the total weight of the system is conserved.

Split balls are expressed in multiplicative forms and thus constitute pro-Benford influences in the system. Consolidated balls are expressed in additive form as $(\text{Ball}_1 + \text{Ball}_2)$ and thus constitute anti-Benford and pro-Normal influences. Yet, if at a later cycle a consolidated ball is randomly chosen for fragmentation, then the new parts are expressed as $(\text{Ball}_1 + \text{Ball}_2) \times \text{Uniform}(0,1)$ and $(\text{Ball}_1 + \text{Ball}_2) \times (1 - \text{Uniform}(0,1))$, thus involving also multiplicative pro-Benford influences, and so addition and multiplication are mixed here. On the other hand, if at that later cycle, the consolidated ball of $(\text{Ball}_1 + \text{Ball}_2)$ is chosen by chance for another consolidation instead of fragmentation, to be further fused with say Ball #5, then it leads to a newly created ball, which is written as $(\text{Ball}_1 + \text{Ball}_2 + \text{Ball}_5)$, further supporting anti-Benford and pro-Normal influences in the system, and even more strongly so now. Surely, multiplication and addition in this process are on a collision course, readying themselves for a long war of attrition.

The evolution of the arithmetical expressions here is of a random nature, as it all depends on the details and trajectory of the random selections of balls. Empirical exploration of the randomly generated algebraic expressions coming out of the mathematical model via computer simulations, recording and summarizing (in the abstract) tens of thousands of possible scenarios of these algebraic operations emerging from numerous ball selections, points to a tug of war between addition and multiplication, where, on average, approximately one-third of the final expressions of the resultant weights of the balls are additive, and approximately two-thirds of them are multiplicative — and this result springs directly from the ratio of one addition and two multiplications per one full cycle, as in the definition of the process.

Monte Carlo computer simulations of actual values and digits reveal that multiplication decisively triumphs over addition after sufficiently many cycles are performed, with a decisive digital convergence to Benford. Why? Because the process uses the high-OOM variable of $\text{Uniform}(0,1)$, and since the emergence of some skewness is guaranteed via plenty of multiplications, it follows that the process encounters the Achilles' heel of the CLT, and therefore addition is not effective at all in disturbing the multiplicative convergence to Benford, as it pushes too mildly and too slowly toward the Normal. The other basic disadvantage of addition is that it occurs only once in each cycle, as compared with multiplication which occurs twice in each cycle, thus rendering multiplication (with two-thirds) naturally more influential and prevalent than addition (with one-third).

Monte Carlo computer simulations show that after plenty of cycles, denoted by C, the final set of the weights of the balls is quantitatively skewed and nearly perfectly Benford, assuming that $C > (2) \times (L)$ approximately and given the existence of a sufficiently large number of balls in the system such that $L > 300$ approximately. Processes with fewer balls or fewer cycles do not converge fully to Benford, while those with exceedingly low values of C or L are not Benford at all.

Chapter 25

Random Exponential Growth Leads to Positive Skewness and Benford

Exponential growth series are expressed as $\{B, BF, BF^2, BF^3, \ldots, BF^N\}$, where B is the initial base value, N is the number of growth periods, $P\%$ is the constant percent growth rate, and F is the constant multiplicative factor related to the growth rate, as in $F = (1 + P/100)$. For example, for 5% exponential growth with an initial B quantity of 300 as the base, $F = (1 + 5/100) = 1.05$, the series as a data set is {300.0, 315.0, 330.8, 347.3, 364.7, 382.9, 402.0, 422.1, etc.}. Since the arithmetic model of exponential growth is intimately connected with multiplicative operations, as it is presented as repeated multiplications and powers, Benford behavior and quantitative skewness are expected. Indeed, the vast majority of exponential growth series are Benford, with only a few exceptions, but for some series, the Benford configuration is found only if plenty of elements of growth (periods) are considered, not just a few. All exponential growth series, without a single exception, are quantitatively skewed, where the small decisively outnumbers the big. Nevertheless, in practical terms, applicability to Benford's Law here is very rare in real-life data. One rare example is given with the case of end-of-year account balance readings of a frozen bank account untouched for 50 years and without any deposits or withdrawals, so as not to disturb the digital configuration. This data set consists of a long sequence of records of the form {balance end of year 1, balance end of year 2, balance end of year 3, etc., balance end of year N}. Another more realistic example perhaps might be a whole log of hourly bacteria readings for constant bacterial growth at a bio laboratory, steadily providing glucose and water, and recorded for a few consecutive days or several weeks. Very few other real-life examples exist; hence, exponential growth series have very limited manifestations of Benford's Law in actual data.

The above discussion regarding any particular deterministic exponential growth series with a fixed and predictable growth rate involves the consideration of all the quantities during all the time periods $\{1, 2, 3, \ldots, N\}$ as a singular data set. An innovative explanation for the prevalence of Benford's Law in several real-life physical systems was proposed by Kenneth Ross, where he considers the collection of only the last term $\{BF^N\}$ of numerous and distinct exponential growth series, each having a random growth factor, F, and random base, B, but with a fixed number of periods, N, common to all. For each growth series, once the growth factor F has been randomly chosen, it then becomes a singular, fixed, and constant growth factor serving all its periods.

Ross chooses B and F as in the continuous Uniform on $(1, 10)$, where all digits from 1 to 9 enjoy equitable proportions on the interval, arguing that this would supposedly prevent *a priori* any possible bias against any particular set of digits. This choice, though, implies the odd possibility of super-rapid growth, where F could reach very high values close to 10, which is 900%, as in $F = (1 + 900/100) = (1 + 9) = (10)$. Ross' model can be summarized as $(\text{Uniform}_1) \times ((\text{Uniform}_2)^N)$, where each element in his model springs from two independent realizations of these Uniforms and is inserted into the above format. His model then considers numerous such elements as one big data set, which is indeed Benford.

Actual examples of simulated elements from Ross' model are displayed, assuming N is 88:

$$\{3.7 \times (5.6)^{88}, 8.2 \times (2.3)^{88}, 1.3 \times (9.7)^{88}, 6.4 \times (3.5)^{88}, 5.3 \times (1.9)^{88},$$
$$9.8 \times (6.3)^{88}, 2.1 \times (4.2)^{88}, \ldots, \text{etc.}\}$$

This author is proposing a similar model in the spirit of Ross' model, albeit with some extra randomness, namely the collection of only the last term $\{BF^N\}$ of numerous and distinct exponential growth series having a random initial base B, random growth factor F, and random time span N, so that the length of these series is also random. In addition, the factor F is not allowed to vary excessively, from 0% to 900%, but rather more mildly and more realistically, from 0% to 14% for example. Such a scheme could serve as an excellent model for populations of cities and towns, representing the census population data snapshot at a specific moment in time of a large collection of growing cities in a big country, recording and considering only the final (current) quantity of population, and excluding past historical population records. This model would be based on the random initial number of settlers (say 10 people at most), distinct random growth rates for each city (say between 0% and 14%, but fixed for the entire city history once chosen), and of varying random age so that some are old and established cities (say 100 years at most), while some are more recent and newer modern cities, and with inter-migration strictly prohibited in the model.

Denoting the discrete **Uniform{integers}** as opposed to the continuous **Uniform(min, max)**, this model can be written succinctly as

$$\text{Uniform}\{1, 2, 3, \ldots, 10\} \times \text{Uniform}(1.00, 1.14)^{\textbf{UNIFORM}\{\textbf{1,2,3,...,100}\}}$$

Actual examples of simulated cities are given as follows:

$$\{7 \times (1.12)^{63}, \; 4 \times (1.05)^{98}, \; 10 \times (1.01)^{6}, \; 5 \times (1.13)^{22}, \; 2 \times (1.14)^{55},$$
$$1 \times (1.08)^{11}, \; 6 \times (1.07)^{28}, \ldots, \text{etc.}\}$$

Each city has its own unique combination of B, F, and N, although its growth rate F is fixed and identical for all its growth periods once randomly decided upon. Monte Carlo simulations of the model with 25,000 realizations (cities) yielded $\{31.8, 17.0, 11.4, 8.8, 7.9, 6.7, 6.0, 5.5, 4.8\}$ as the first-digit distribution, with an SSD value of 5.4, indicating strong conformity to Benford. For simplicity and in extreme generality, the model uses fractional values, such as 17.85 persons.

Introducing even more randomness into the model can be achieved by letting each city determine its own yearly growth rate F anew randomly, so that for each city and for each year, a distinct $P\%$ of growth is chosen randomly anew between 0% and 14%, each 1st January.

Three actual examples of three simulated cities, with 11, 7, and 9 years of growth, are displayed as follows:

{8 × (1.11) × (1.04) × (1.11) × (1.09) × (1.14) × (1.02) × (1.07) × (1.02) × (1.01) × (1.13) × (1.04),

5 × (1.07) × (1.01) × (1.07) × (1.10) × (1.12) × (1.05) × (1.08),

2 × (1.03) × (1.05) × (1.13) × (1.01) × (1.07) × (1.06) × (1.12) × (1.10) × (1.02),

and so on . . .}

Theoretically, such a model involving significantly more randomness should yield digit distribution even closer to Benford. Indeed, this is the case, and Monte Carlo simulations with 25,000 realizations (cities) yielded {30.6, 17.9, 12.4, 9.5, 8.0, 6.5, 5.8, 4.9, 4.5} as the first-digit distribution, with an SSD value of 0.4, indicating a near-perfect conformity to Benford.

Models of the last term of numerous exponential growth series, where both the base B and the factor F are fixed as constants, being identical for all cities, while only the number of N periods or years varies randomly (and widely) by city, also converge decisively to Benford. Models of the last term of cities with an identical fixed age N as well as an identical fixed initial population base B but with distinct random growth factors F (of sufficient variability) also converge strongly to Benford. All these increase the scope and applicability of the generic last-term exponential growth model to additional types of real-life physical data. In extreme generality, the enormous significance of all such last-term random growth schemes is that they can serve as Benford models for all entities that spring into being gradually via random growth, be it a set of cities and towns with growing populations, rivers forming and enlarging gradually along incredibly long geological time scales, biological entities growing or forming, stars and planets forming, and so forth. The potential scope covered by this result is enormous!

For the rest of the book and in all subsequent chapters, the term "exponential growth series" reverts back to its original meaning, with deterministic, known, and fixed parameters, of a predictable and unique growth rate, and of just a single series, involving the consideration of all its quantities during all the time periods as a singular data set, and not that of Ross' model.

Chapter 26

Data Aggregation Leads to Positive Skewness and Often to Benford

It may not be obvious, but surprisingly fairly often, real-life data sets consist of numerous, smaller, and "more elemental" mini sub-sets, all of which are aggregated and fused together to generate the final and apparent data on hand. In addition, it can be demonstrated in general that appending various data sets into a singular and much larger data set leads to quantitative skewness in favor of the small and toward Benford, assuming that these data sets commonly start from a very low value, ideally such as 0 or 1, and that they terminate at highly differentiated endpoints, so that some span short intervals while others span longer intervals, as is typically the case for data creation by way of data aggregation. Let us demonstrate this quantitative tendency in data aggregation by combining the following six imaginary data sets:

Data Set A: {2, 3, 5, 7}
Data Set B: {1, 4, 6, 9, 13, 14}
Data Set C: {2, 6, 7, 9, 11, 15, 16, 21}
Data Set D: {1, 2, 6, 8, 13, 14, 19, 23, 25}
Data Set E: {3, 4, 8, 12, 15, 19, 22, 24, 29, 35, 41}
Data Set F: {1, 5, 8, 11, 12, 17, 19, 24, 27, 32, 38, 43, 47}

The following vector depicts the combined data set of A, B, C, D, E, and F, ordered from low to high:

$\{1, 1, 1, 2, 2, 2, 3, 3, 4, 4, 5, 5, 6, 6, 6, 7, 7, 8, 8, 8, 9, 9, 11, 11, 12, 12, 13, 13,$
$14, 14, 15, 15, 16, 17, 19, 19, 19, 21,\ 22,\ 23,\ 24,\ 24, 25, 27, 29, 32, 35, 38,$
$41, 43, 47\}$

Figure 20 depicts these six data sets superimposed, thus allowing us to visualize the concentration and piling up of numerous small values that occur on the left, in sharp contrast to the diluted spread of only a few big values that occur on the right.

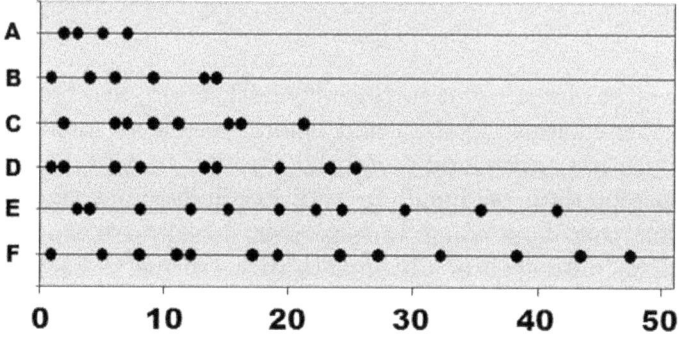

Figure 20.   Visualizing the Creation of Many Small but Only a Few Big Values

It is important to note that, in spite of the fact that each component data set A–F is quantitatively structured in an even and balanced manner approximately and very roughly in the spirit of the Uniform distribution, where no size is significantly preferred over any other size, yet, for the aggregated data set, the small decisively outnumbers the big. The dynamics behind such a tendency to produce more small quantities than bigger ones is the differentiated overlapping of ranges for the aggregated data set. Overlapping here occurs more on the left for small values and less so on the right for big values.

Let us illustrate this more vividly by listing house numbers for several streets pertaining to a hypothetical mail address data set of an imaginary post office branch in one factitious small town, which may typically be listed as follows:

$\{1, 2, 3, 4, 5, 6\}$ — Floral Drive
$\{1, 2, 3, 4, 5, 6, 7, 8, 9, 10, 11, 12, 13, 14, 15\}$ — Pine Avenue
$\{1, 2, 3, 4, 5, 6, 7, 8, 9, 10, 11, 12, 13, 14, 15, 16, 17, 18, 19\}$ — Main Street
$\{1, 2, 3, 4, 5, 6, 7, 8, 9, 10, 11, 12\}$ — South Street
$\{1, 2, 3, 4, 5, 6, 7, 8, 9\}$ — Lodge Street

All streets necessarily start at house number 1, namely the first house on the left or right corner, but each street terminates at a different house number depending on the length of the street; hence, many numbers pile up on the left range of small values, leaving the right range of big values diluted and rarer. The aggregation of all house numbers from all the streets in a given city constitutes an implicit data aggregation, resulting in a positive skewness in favor of the small.

Not all data aggregations lead to Benford exactly, but most are fairly close, and at a minimum, their first-digit distribution resembles the Benford configuration. As one concrete real-life example, address data pertaining to Prince Edward Island in Eastern Canada containing 23,633 addresses, are downloaded, digits are counted, and the data set is verified to comply with Benford's Law. The link to the website is http://www.gov.pe.ca/civicaddress/download/. Clearly, the focus here is not on the street number or the zip code but exclusively on the house number. The first-digit distribution is $\{30.9, 18.5, 15.1, 10.0, 6.1, 6.0, 5.2, 4.3, 4.0\}$, and the low SSD value of 13.1 indicates that this house number data is fairly close to the Benford configuration.

This author has performed extensive computer calculations in relation to data aggregations in the abstract, using generic models to aggregate sets of consecutive integers with specified and exact lower and upper bounds, in order to calculate the resultant first-digit distributions. A detailed account of this project can be found in the author's 2014 Benford book, in Chapter 50 tilted "A Leading Digits Parable", Chapter 51 titled "Simple Averaging Scheme as a Model for Typical Data", and Chapter 52 titled "More Complex Averaging Schemes".

The first simple model with three parameters is coined "**simple averaging scheme**", and it investigates the first digits for multiple intervals, made exclusively of the integers lying along the x-axis, all starting from some common lower bound, LB, which is typically 1, and having differentiated length, as the intervals are being made progressively longer by systematically increasing their upper bound, UB, by one integer at a time, from the initial UB-min to the final UB-max. The plan is then to obtain an aggregated digital distribution representing all the intervals by simply taking the (non-weighted) average of the digital distributions of all the intervals, interval by interval, as opposed to the average first digits of the totality of all the integers themselves within all the intervals. For parameters $LB = 1$, UB-$min = 3$, and UB-$max = 33$, there are 31 intervals, namely $\{1, 2, 3\}$, $\{1, 2, 3, 4\}$, $\{1, 2, 3, 4, 5\}, \ldots, \{1, 2, 3, 4, 5, 6, 7, \ldots, 31, 32, 33\}$. For parameters $LB = 1$, $UB - min = 1$, and $UB - max = 9$, there are 9 intervals: $\{1\}$, $\{1, 2\}$, $\{1, 2, 3\}$, $\{1, 2, 3, 4\}$, $\{1, 2, 3, 4, 5\}$, $\{1, 2, 3, 4, 5, 6\}$, $\{1, 2, 3, 4, 5, 6, 7\}$, $\{1, 2, 3, 4, 5, 6, 7, 8\}$, and $\{1, 2, 3, 4, 5, 6, 7, 8, 9\}$, and the average of the first-digit configurations of the 9 intervals is $\{31.4,$ 20.3, 14.8, 11.1, 8.3, 6.1, 4.2, 2.6, 1.2$\}$. Somewhat different results are obtained for the much wider and more numerous set of intervals with parameters $LB = 1$, $UB - min = 1$, and $UB - max = 10,000$, where the average of the first digits comes out as $\{24.2, 18.3, 14.5,$ 11.7. 9.5, 7.6, 6.0, 4.6, 3.4$\}$. In general, this is an excellent model for **city** post office address data on house numbers, with LB equal to 1, UB-min equal to the number of houses for the shortest street in the city, and UB-max equal to the number of houses for the longest street in the city.

There is no hope of obtaining the Benford configuration here, unless we go on to the next level of complexity in averaging out multiple such simple averaging schemes themselves while gradually varying the x-axis focus of each simple scheme. This is typically done by letting the lower bound LB stay fixed at 1 and letting the upper bound start at a fixed point, UB-min, as well, while allowing the upper bound to terminate at a variety of different locations between UB-max-$lowest$ and UB-max-$highest$. For example, $LB = 1$, $UB - min = 10$, $UB - max - lowest = 100$, and $UB - max - highest = 1000$ refer to multiple simple averaging schemes where LB is fixed at 1, UB-min is fixed at 10, while

UB-max varies from 100 to 1000. In other words, this involves the consideration of multiple simple averaging schemes with LB fixed at 1, while UB varies on the intervals (10 to 100), (10 to 101), (10 to 102), ..., (10 to 998), (10 to 999), (10 to 1000), which is then followed by the averaging of all these 901 different simple averages. The first-digit result here comes out as {30.4, 18.9, 13.1, 9.8, 7.6, 6.2, 5.2, 4.6, 4.1}, which is closer to Benford when compared with the result from any simple averaging scheme. This more complex scheme with four parameters is termed the "**average of averaging scheme**", and it is appropriate for a **country** with many cities in need of averaging out their house numbers in the address data, so that a country-wide resultant first-digit configuration is obtained.

The next higher-order scheme involves averaging the averages of the averages, with parameters *LB*, *UB-min*, *UB-max-lowest*, *UB-max-highest-starting*, and *UB-max-highest-ending*. In one example, with parameters 1, 30, 60, 250, and 500, respectively, the first digits came out as {29.3, 16.4, 12.6, 10.1, 8.4, 7.1, 6.1, 5.3, 4.7}, which is even closer to Benford. This even more complex scheme with five parameters is termed the "**average of averages of averaging schemes**", which is appropriate for the **global** address house data, fusing and averaging out house number data from all the countries in the world.

The same rationale that propelled us to progress from city to country and then globally necessitates the continuation of averaging schemes of even higher orders. Indeed, convergence to Benford is rapidly achieved here, while results become completely independent of the values chosen for all the LB and UB parameters, and this crucial feature lends more credibility and robustness to the whole averaging scheme theory in general. In other words, the scheme is totally independent of parameters in the limit as higher and higher averaging orders are employed.

This author independently proposed such an averaging approach in Benford's Law in 2005. It was later realized that long before that, in 1963, the late Betty Flehinger presented a rigorous mathematical proof that an iterated averaging scheme for the integers on the positive number line approaches Benford in the limit as the number of such iterations goes to infinity. The above discussion and the integer aggregation project could serve as the conceptual framework for going in such a roundabout way in calculations and then having the Benford configuration emerge; and Flehinger's abstract iterative scheme could also refer to a concrete usage of numbers in the real world, not merely to the number line itself. Flehinger's algorithm in the abstract could not amount to a real-life explanation of the law, even though one might passionately argue that her proof has its strong basis in none other than the number line itself! The author wishes to convey his strong sense of affinity and rapport with Flehinger for thinking along almost the same lines in the quest for an explanation of Benford's Law.

Chapter 27

Chains of Statistical Distributions and Benford's Law

Two features guarantee quantitative skewness in favor of the small in data aggregation, such as the one depicted in Figure 20. The first feature is the approximate common beginning of the minimum for all the component data sets, which are fixed at around 0, 1, or some low value. The second feature is the uncertain and highly varied (i.e., random) termination at the end for the maximum, namely that each component data set has its own particular maximum. Such a skewed quantitative tendency or mechanism in aggregations of naturally occurring and elemental real-life data is one of the main causes and explanations for the Benford's Law phenomenon in the real world. Formalism in mathematical statistics draws inspiration from such types of data aggregation and points to a slightly different yet similar abstract process or model, termed the "chain of two Uniform distributions", where parameter a of the Uniform is fixed at min, while parameter b is uncertain and random, as it varies and is obtained via another Uniform distribution, namely the statistical chain **Uniform(min, Uniform(maxA, maxB))**. The parameters min, maxA, and maxB are of fixed values, namely constant numbers, and min $\leq$ maxA $<$ maxB. The chain could vaguely be interpreted as the superimposition or the union of a large set of uniforms with distinct parameter b values but a common parameter a value, while each Uniform is granted equal weight and equal importance within the entire model, in spite of having distinct lengths (namely different maximums). As a very rough and approximate continuous model for the discrete data sets A, B, C, D, E, and F, shown in Figure 20, the chain **Uniform(1, Uniform(7, 47))** might be used to generate generic values of similar configuration, and this is especially so in light of the fact that each data set A–F was made roughly in an even and balanced manner and in the spirit of the Uniform.

The chain **Uniform(0, Uniform(0, B))** converges only to an approximate Benford configuration, with SSD values fluctuating between 25 and 57, depending on the exact value of B. This chain hints at the simple averaging scheme (city), except that it is of the infinite continuous format and not of the finite discrete format involving only integers. Two or three additional uniforms need to be added here to the chain, enlarging the number of its sequences and thus its length, in order to increase uncertainty and randomness, and hopefully leading to a forceful convergence to Benford. Indeed, when four Uniform distributions are chained with regard to parameter b only and with the ultimate b value set at any arbitrary value, say 55, while parameter a is fixed at 0 for all four Uniform distributions, a near-perfect convergence to Benford occurs. Formally, the entire arrangement of this chain scheme is written as **Uniform(0, Uniform(0, Uniform(0, Uniform(0, 55))))**. This four-sequence chain process can also be described as a step-by-step simulation scheme as follows: simulate a single value from Uniform(0, 55) and call it R; then simulate a single value from Uniform(0, R) and call it G; then simulate a single value from Uniform(0, G) and call it P; then simulate a single value from Uniform(0, P) and call it Q. This last Q value is the final result of the entire chain. In one Monte Carlo run, with 10,000 simulated values, the first digits came out as {30.6, 17.5, 11.8, 9.3, 8.4, 7.0, 5.9, 5.1, 4.5}, and its extremely low SSD value of 1.3 indicates that this chain is nearly perfectly Benford. The choice of 55 is totally arbitrary here for such a long chain of four sequences, and other values for the ultimate value of parameter b yield nearly identical results and the same forceful convergence to Benford. This longer chain hints at the next higher-order level from the average of averages of averaging schemes (global +1), except that it is of the infinite continuous format and not of the finite discrete format involving only integers. Empirical examinations of resultant digit configurations show that the length of the chain (i.e., the number of Uniform sequences involved) correlates positively with closeness to Benford. Hence, the longer the chain, i.e., the larger the number of sequences of tied-up Uniforms, the closer we get to $LOG_{10}(1 + 1/d)$, and this is so conceptually as randomness increases with each added sequence.

Other types of chains of distributions, applying a variety of distribution forms (not merely the Uniform), especially with a large number of dependent sequences of over three or four, tying up location and/or scale parameters, yield quantitative skewness and are almost always nearly perfectly Benford. This lends the chain-of-distributions feature a truly colossal scope of manifestations, occurrences, and applications in the context of quantitative skewness in general and Benford's Law in particular. Figure 21 depicts one such possible complex chain of four sequences. The last row at the bottom contains traditional or orthodox distributions with fixed parameters, terminating all dependent relationships.

Defining $Q = 7$ means that we are absolutely certain about the value of Q. We are less certain about which face of the dice will show up during a game in the casino, but we know that the possibilities are only $\{1, 2, 3, 4, 5, 6\}$ and certainly not 7 or 368. Determining or predicting one realization from one orthodox and traditional distribution with fixed and known parameters is more difficult and more random, yet it's not totally uncertain. For example, the variable Normal(5, 1) is somewhere between 2 and 8, but mostly around 5. The variable Uniform(2, 7) is certainly not below 2, nor above 7. In contrast, obtaining a realization from a Uniform with min parameter a fixed at 3 but with max parameter b itself being uncertain as it is derived randomly from Uniform(9, 11) involves more randomness and uncertainties, and this is the reason why the chain Uniform(3, Uniform(9, 11)) is considered "**more uncertain**", and the chain process could be phrased as "**randomness within randomness**". Moreover, determining or predicting one realization from the long and complex chain in Figure 21 is by far more challenging and difficult, since the values involved are much more random and uncertain. In this sense, the decisive convergence to Benford of nearly all long and complex chains of distributions leads us to view **Benford's Law** also as "**the property of the super random number**".

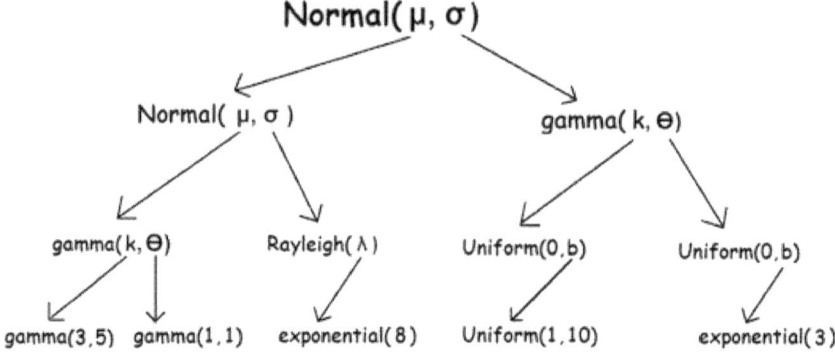

Figure 21.   Pyramid-Like Arrangement of Complex Distributions Chain

As another example with a chain of three Exponential distributions, **Exponential(Exponential(Exponential(7)))** converges to Benford nearly perfectly. In 10,000 simulation runs, first digits came out as {30.3, 18.1, 12.3, 10.2, 8.0, 6.5, 5.1, 5.0, 4.4}, and the extremely low SSD value of 1.2 here indicates that this chain is nearly perfectly Benford. Since the Exponential itself is somewhat close to Benford, convergence here is more rapidly achieved.

As an example of a hybrid chain composed of three distinct distributions, **Exponential(Normal(Uniform(15, 21), 3))** converges to Benford as expected. In 10,000 simulation runs, the first digits came out as {29.6, 16.0, 11.9, 10.0, 8.5, 7.4, 6.1, 5.6, 4.8}, and the very low SSD value of 4.5 here indicates that this chain is very close to Benford.

The author's first conjecture is that an infinitely long chain of distributions should obey Benford's Law exactly. The rationale behind it is that an infinite sequence represents "infinite uncertainty" and "infinite randomness". This infinite chain hints at the infinitely iterated averaging scheme for the integers on the positive number line and at Flehinger's proof. Yet, the initial investigation of the chain phenomenon by this author revealed that although the vast majority of whatever distributions and types of parameters involved in chaining are Benford, there exists a tiny minority which adamantly refuses to obey the law of Benford. All these were quite perplexing and mysterious. In order to understand the principle involved, it was necessary to construct a large table containing numerous distributions and parameters, showing their Benfordian or non-Benfordian behavior under chaining. Fortunately, intuition or instinct suggested adding the algebraic expressions for the average and the median as a function of parameters, and this helped in discerning a pattern which would hint at the criteria for compliance with the law. That attempt indeed led to a decisive success. Interested readers are referred to Chapters 102 and 103 of the author's 2014 book on Benford for details. The conclusion of that entire investigation is summarized here in the following paragraphs.

Scale parameters, such as λX or X/λ (divisions and multiplications), as well as location parameters, such as $X - \mu$ (subtractions), usually respond vigorously to chaining, prefer the small over the big, and obey Benford's Law (i.e., chainable). Shape parameters, such as X^k (powers), usually do not respond to chaining at all, show no preference for the big, small, or any size, and disobey Benford's Law (i.e., not chainable).

More precisely, an *ineffective parameter* that does not continuously involve itself in the expression of centrality (such as the mean, median, or midpoint) is not chainable at all, while an *influential parameter* that does continuously involve itself in the expression of centrality is indeed chainable.

The phrase "not continuously involved" means that the partial derivative $\partial(\text{center})/\partial(\text{parameter})$ goes to 0 in the limit of high values of the parameter, namely that centrality, such as the average, median, or midpoint, is not affected as the parameter is further increased, and that beyond the initial low parametrical values, the parameter does not sway centrality anymore.

$$\lim_{\text{parameter}} \to \infty \quad \partial(\text{center})/\partial(\text{parameter}) = 0$$

Hence, if a parameter does not play any role in the determination of centrality beyond the initial few low values, then it's not chainable at all. The consequence of the above finding is as follows: since nearly all scale and location parameters (always and continuously) play a significant role in centrality, they are generally chainable; however, since most shape parameters do not play any role in centrality, they are generally not chainable.

For example, the chain Uniform(0, Uniform(0, b)) utilizes parameter b, which always affects the centrality of the inner distribution Uniform(0, b) since Expected(X) = $b/2$, which strongly and continuously involves b, or, equivalently, that the partial derivative with respect to b is $1/2$ in the limit as b goes to infinity, as it is never diminishing, hence the chain is close to Benford.

For the Wald distribution, with location parameter μ and shape parameter λ, the average is expressed as $\text{Exp}(X) = \mu$, namely that the average involves only the location parameter μ but not the shape parameter λ, or, equivalently, that the partial derivative with respect to μ is 1 in the limit as μ goes to infinity, never diminishing, and that the partial derivative with respect to λ is 0 in the limit as λ goes to infinity; therefore, the location parameter μ is indeed decisively chainable, as verified via computer simulations, while the shape parameter λ is not chainable at all, as verified via computer simulations.

For the Pareto distribution with probability density function $\text{PDF} = ab^a/x^{(a+1)}$, shape parameter $a > 1$, and scale parameter $b > 0$, the average is expressed as $\text{Exp}(X) = b(a/(a-1))$.

In the limit as a goes to infinity, the average is $b(1)$, or simply b, not involving a at all, or, equivalently, the partial derivative with respect to a is 0 in the limit as a goes to infinity. Also, in the limit as b goes to infinity, the average never diminishes, or, equivalently, the partial derivative with respect to b is $a/(a-1)$ in the limit as b goes to infinity, never diminishing. As a consequence, the scale parameter b is decisively chainable, as verified via computer simulations, while the shape parameter a is not chainable at all, as verified via computer simulations.

In order to understand conceptually the dichotomy between influential parameters and ineffective parameters, it is first necessary to recall the vague interpretation mentioned earlier of a short two-sequence chain $G(R$ defined on the range from p to q) as the superimposition or union of a large set of G distributions with parameter varying from p to q, as in the random way of the R distribution. Now, we are ready to answer the question of why only those parameters that are decisively and continuously involved in the expression of centrality should be chainable. The answer is quite straightforward, because the mere act of chaining in any way such an influential parameter yields, at a minimum, a set of distributions with increasing and stretching upper bounds (maximum), derived from the fact that the parameter significantly affects location and centrality, while hopefully with lower bounds (minimum) staying fixed near 0, 1, or any other low number, yielding an overall aggregated or superimposed density with a one-sided tail to the right, being positively skewed. Nothing of the sort could happen if the ineffective parameter to be varied by chaining completely stops affecting centrality beyond its few initial low values.

In addition, the author's second chain of distribution conjecture predicts the manifestation of Benford's Law exactly for short chains of distributions, even with just two sequences, assuming the chain uses a distribution which obeys Benford's Law exactly for the innermost parameter in its ultimate sequence. For example, the short chain Uniform(0, Exponential Growth Series) is predicted to closely obey Benford's Law, even though it only has two sequences. This is so since an Exponential Growth Series in and of itself obeys Benford's Law, assuming a sufficient number of elements. In another example, the chain Uniform(0, Lognormal(5, 2)) is predicted to be nearly perfectly Benford, even though it only has two sequences, and this is so because the high value of the shape parameter, being well over 1.2, renders the Lognormal itself nearly perfectly Benford.

In extreme generality, and concisely in symbols, the second conjecture states that

Any Distribution(Any Benford) = Benford

A related extrapolation of the second conjecture states that, with each new added sequence (elongating the chain), the chain evolves and gains skewness, as well as becoming a notch closer to the Benford digital configuration. In other words, within a long sequence of distributions forming a chain, as we march outward from the innermost distribution toward the outermost distribution, we encounter an increasing measure of Benfordness along the way, and that Benfordness never diminishes, not even temporarily. For example, the extrapolation states that for the short chain $G(R(S(\text{fixed parameter})))$, deviation from Benford decreases steadily so that $\{\text{SSD of } R(S(\text{fixed parameter}))\} < \{\text{SSD of } S(\text{fixed parameter})\}$. In the same spirit, the extrapolation also states that $\{\text{SSD of } G(R(S(\text{fixed parameter})))\} < \{\text{SSD of } R(S \text{ (fixed parameter)})\}$. For a short chain of only two sequences, the extrapolation of the second conjecture implies that if the inner random parameter is exactly Benford, then the chain applying it must be at least as Benford, namely Benford! Hence, the extrapolation of the second conjecture implies the second conjecture!

The relevance of the chain-of-distributions model to real-life data becomes more apparent when considering the causality in life and nature, the interconnectedness in the world, and the dependencies of some entities upon other entities. Exact cause-and-effect relationships in the deterministic realms, such as in physics, chemistry, astronomy, geology, and biology, often lead to skewed results and scenarios, which can be modeled as chains of distributions, especially when the phenomenon is aggregated and examined on a large scale. For example, lengths, widths, and water volume discharge of rivers all depend on average rainfall (being the parameter), and rainfall in turn depends on sunspots, prevailing winds, and geographical location, all serving as parameters of rainfall. Health indexes of people may depend on overall childhood nutrition, while nutrition in turn may depend on overall economic activity, which in turn depends on economic policy, geopolitical stability, weather-related events such as droughts and flooding, and so forth. All these lead to the conclusion that, often, some physical measurements serve as parameters for other physical measurements. This is another chief cause and explanation for why so often the Benford digital configuration is found in real-life data.

The mathematician Steven J. Miller gave a rigorous mathematical proof for the author's first and second chain conjectures for three particular one-parameter distribution cases, where the probability density function is defined exclusively on the positive infinity range of $(0, +\infty)$, namely for Uniform$(0, b)$, exponential(ρ), and $|$Normal$(0, \sigma)|$. Computer simulations and conceptual reasoning, however, clearly point to further applicability, far more so than in Miller's relatively restricted cases. The author is greatly thankful to Miller for his work.

Chapter 28

Meta-Explanation or the Explanation of all Explanations

We note with amazement the remarkable versatility and prevalence of Benford's Law in the natural world as well as in the realm of its purely abstract mathematical and statistical manifestations. All roads (almost) lead to Rome. In the same vein, the following actions all lead to Benford, namely: break it up, glue stars and planets together, multiply it, divide it, aggregate many data sets together, let it grow randomly, chain distributions up, spread it around (populations and such), shake it up (earthquakes), carve rivers out of mountains, and so forth. Surely the nearly universal tendency in all these processes and actions is the resultant positive skewness, where the small is numerous and the big is rare, and this is apparent, expected, and intuitive, as narrated in the chapters of this section, yet what is surprising here is that we always arrive at the same identical numerical measure indicating how much the small is more numerous than the big, namely $\text{LOG}(1 + 1/d)$, for all these diverse and seemingly unrelated scenarios and processes.

This is a very curious state of affairs; seemingly totally distinct and unrelated processes lead to the same numerical results! Imagine seven chefs, French, Thai, Russian, Chinese, Japanese, Indian, and South African, on different continents, using totally different ingredients, cooking styles, and recipes; yet they all come up with seven dishes equal in taste, consistency, flavor, smell, and colors! In the same vein, it appears far-fetched that totally different physical processes and abstract models yield nearly exact digital distribution for the nine digits! Surely, there must be some fundamental commonality and correspondence between all of these natural processes and physical arrangements!

The ultimate quest in Benford's Law is to somehow unite all these diverse physical processes and mathematical explanations or models into a singular concept and show that that fundamental concept is Benford, although such a goal may prove elusive. Surely Mother Nature would be immensely upset upon learning that someone on this minor and insignificant planet is attempting to simplify and belittle her complex and versatile digital behavior. Mathematicians wishing to take on this challenge should be warned: she is quite malicious and vindictive when provoked!

This author would reluctantly and very hesitantly suggest as a meta-explanation the curvy-closure property of the histogram of related logarithms (to be discussed later in Chapter 32), together with the typical high variability of data, where the OOM is greater than 3 and which together nearly always appear, decisively and consistently so, not only in real-life empirical data but also in all of the various abstract processes, theoretical explanations, and mathematical models, leading to the Benford configuration.

Yet, this meta-explanation, in essence, is a circular argument of sorts. We are merely kicking the can down the road for temporary relief, and instead of wondering why $LOG(1+1/d)$ appears almost everywhere for the first digits, we are now wondering why the curvy-closure property appears almost everywhere in the histograms of logarithms of numbers in data, and why OOM is nearly always so high. A somewhat superior meta-explanation may be offered perhaps via the postulate on relative quantities (to be discussed later in Chapter 48), while extrapolating and interpreting the postulate to stand as an idealized model (of the k/x distribution type) in the aggregate, overall, via the *global* perspective of the data structure in real-life data, but which is nearly always not of the k/x distribution format *locally* on the left or right parts of the data.

SECTION III

The Logarithmic Perspective

Chapter 29

Benford's Law as Uniformity of Mantissa

The idea of the logarithm is closely related to the basic concepts of our number system. Any number expressed in our number system is a particular linear combination of 10^{INTEGERS}, with the restriction on coefficients being integers and only in the set $\{0, 1, 2, 3, 4, 5, 6, 7, 8, 9\} < 10$.

For example, the number $8745 \equiv (8) \times (10^3) + (7) \times (10^2) + (4) \times (10^1) + (5) \times (10^0)$. Asking about the value of the logarithm of 8745 is an attempt to forcefully condense and squeeze all these distinct integer powers of 10 into a singular power of 10, come what may, while further restricting its coefficient to be 1. For this to work out, though, we must be flexible and allow that singular power to be a non-integer fraction. Hence, the question of the value of L in $\text{Log}_{10}(8745) = L$ translates into the following equation with unknown L: $(8) \times (10^3) + (7) \times (10^2) + (4) \times (10^1) + (5) \times (10^0) \equiv (1) \times (10^L)$. The unique solution here is $L = 3.9418$, or, in other words, $\text{LOG}_{10}(8745) = 3.9418$, which is therefore $(8) \times (10^3) + (7) \times (10^2) + (4) \times (10^1) + (5) \times (10^0) = (1) \times (10^{3.9418})$. If one objects to the use of our number system in the expression 3.9418 for the logarithm, then this can be remedied by expressing it as the ratio of two quantities, namely 4943/1254, and the logarithm can be written as $(8) \times (10^3) + (7) \times (10^2) + (4) \times (10^1) + (5) \times (10^0) = (1) \times (10^{(4943/1254)})$. Higher precision can be used for more exact values, such as 3.9417598, via the use of larger numerators and denominators — in spite of the fact that logarithms of most numbers are irrational. The meaning of the fractional power of any number Q is the combination of standard integral power and the Nth root, so that $Q^{(\text{NUMERATOR/DENOMIONATOR})} = Q^{(\text{POWER/ROOT})} = (Q^{\text{POWER}})^{(1/\text{ROOT})}$, with $(G)^{(1/N)}$ meaning the Nth root of G. For example, $8^{1/3}$ = the cube root of 8 = 2, and $16^{1/2}$ = the square root of 16 = 4. As another example, $125^{(2/3)} = (125^{1/3})^2 = $ (the cube root of 125)$^2 = 5^2 = 25$.

We might pause for a brief moment and acknowledge that our number system is not only arbitrarily defined and invented, but that it's quite peculiar! Yet, this number system is extremely efficient, or perhaps the most efficient number system in existence. Moreover, it is essential that we acknowledge that the concept of the logarithm itself is even more peculiar than our number system, and this is so in spite of its enormous use and importance in mathematical disciplines and the sciences.

We can now clearly observe and visualize how digits and logarithm gradually grow in unison, steadily increasing from low levels to higher levels. This demonstrates the intimate relationship between the digital configuration of any given number and its logarithmic representation. Readers should focus only on the fractional part of the logarithm to the right of its decimal point, called mantissa, while ignoring the integer part of the logarithm to the left of its decimal point.

$$1487 = (1) \times (10^3) + (4) \times (10^2) + (8) \times (10^1) + (7) \times (10^0)$$

$$= (1) \times (10^{3.1723}) \quad \text{mantissa} = 0.1723$$

$$1488 = (1) \times (10^3) + (4) \times (10^2) + (8) \times (10^1) + (8) \times (10^0)$$

$$= (1) \times (10^{3.1726}) \quad \text{mantissa} = 0.1726$$

$$1512 = (1) \times (10^3) + (5) \times (10^2) + (1) \times (10^1) + (2) \times (10^0)$$

$$= (1) \times (10^{3.1796}) \quad \text{mantissa} = 0.1796$$

$$3686 = (3) \times (10^3) + (6) \times (10^2) + (8) \times (10^1) + (6) \times (10^0)$$

$$= (1) \times (10^{3.5666}) \quad \text{mantissa} = 0.5666$$

$$6711 = (6) \times (10^3) + (7) \times (10^2) + (1) \times (10^1) + (1) \times (10^0)$$
$$= (1) \times (10^{3.8268}) \quad \text{mantissa} = 0.8268$$
$$6712 = (6) \times (10^3) + (7) \times (10^2) + (1) \times (10^1) + (2) \times (10^0)$$
$$= (1) \times (10^{3.8269}) \quad \text{mantissa} = 0.8269$$

In the most simplistic way, the mantissa could be described as the fractional part of the logarithm (base 10) of a number, although this definition is not true for very small numbers less than 1 having negative logarithmic values. For $X < 1$, the mantissa is the complement of the fractional part of the log relative to 1, namely 1 minus the fractional part of the absolute value of $LOG(X)$.

For $X = 870$, Log is $+2.93952$ and mantissa is 0.93952.
For $X = 25.7$, Log is $+1.40993$ and mantissa is 0.40993.
For $X = 0.479$, Log is -0.31966 and mantissa is 0.68034.
For $X = 0.064$, Log is -1.19382 and mantissa is 0.80618.

Let us consider the following four numbers and their corresponding log values:

8.5274 Log $= 0.9308$
85.274 Log $= 1.9308$
852.74 Log $= 2.9308$
8527.4 Log $= 3.9308$

The fractional part of the logarithm, i.e., the mantissa, is the same for these four numbers, namely 0.9308. The digital structure is also the same for these four numbers. The first digit is always 8, the second digit is always 5, the third digit is 2, the fourth digit is 7, and the fifth digit is always 4.

Digital configuration (with all orders considered) and mantissa are basically two distinct ways of indicating or expressing the same numerical condition. Hence, mantissa has a one-to-one correspondence with digital configuration (with all orders considered). Tell me the mantissa, and I will be able to immediately know all the digits composing the number, except that I will not know where the decimal point is. Tell me all the digits composing the number, and regardless of the decimal point, I will be able to immediately know the mantissa.

Integral powers of ten (IPOT), namely $10^{\textbf{INTEGER}}$, with negative, positive, or zero integers, such as 0.01, 0.1, 1, 10, 100, and 1000, derived from 10^{-2}, 10^{-1}, 10^{0}, 10^{1}, 10^{2}, and 10^{3}, respectively, are exceedingly important in the context of Benford's Law since the cycles of the first digits revolve around them, always starting exactly at one IPOT value and completing a full cycle from 1 to 9 by terminating at the next IPOT value. Adjacent integral powers of ten (AIPOT) are two sequential (neighboring) IPOT numbers, or simply the pair $10^{\textbf{INTEGER}}$ and $10^{\textbf{INTEGER}+1}$, such as 1 and 10, 10 and 100, and 100 and 1000. The partitioning of the entire length of the x-axis according to the AIPOT points, such as $[1, 10)$, $[10, 100)$, $[100, 1000)$, and $[1000, 10000)$, is highly significant in terms of first digits as well as mantissa values.

Both first digits and mantissa cycle together in unison around these AIPOT intervals, over and over again. Both start their cycles at some IPOT point, and both terminate their cycles at the next IPOT point, in unison, from their lowest levels to their highest levels, from their tails to their heads.

The mantissa cycles fully from 0 to 1 on each AIPOT interval as it gradually grows from 0 to 1, only to suddenly collapse to 0 again, repeatedly. As the mantissa gradually grows from 0 to 1 on any of these subintervals, the first digits advance in unison from 1, to 2, to 3, and finally to 9, and there is no reversal, no retreat whatsoever. In other words, a higher mantissa value is associated with a higher digital configuration. Higher digital orders, such as second and third orders, also steadily increase without any reversal or retreat on these subintervals as the mantissa grows.

If one points at an imaginary mathematical pin and swings it gradually and smoothly from 10 to 100 along the x-axis, the entire digital repertoire and the manifestations of all the digital orders are encountered in their natural orders from low to high, while just below it, the vague shadow of the movements of the pin projected onto the M-axis of the mantissa swings from 0 to 1 in exact one-to-one correspondence, demonstrating its entire mantissa repertoire accordingly.

The transition from any positive number X to its decimal base 10 logarithm is accomplished by denoting it as $\text{Log}(X)$. The transition from the logarithm back to the number X itself is accomplished by simply taking 10 to the power of $\text{Log}(X)$, hence $X = 10^{\textbf{Log}(X)}$.

Since the logarithm of X can be written and expressed as $\text{Log}(X) = \text{INTEGER} + \text{MANTISSA}$, it follows that $X = 10^{\text{Log}(X)} = 10^{(\text{INTEGER}+\text{MANTISSA})} = (10^{\text{INTEGER}}) \times (10^{\text{MANTISSA}})$.

For example, $10^{3.9308} = 10^{(3+0.9308)} = (10^3) \times (10^{0.9308}) = (1000) \times (8.5274) = 8527.4$.

Therefore, instead of stating in great detail, and with much effort, how all the digital orders are distributed, one might as well consider (more concisely) how their mantissas are distributed, and as a consequence, all digital orders are determined in one fell swoop!

As it happens:

Benford's Law implies uniformity of mantissa.
Uniformity of mantissa implies Benford's Law.

Hence, converting any large Benford data set (with $X > 1$ say) from its normal numbers into a set of logarithm values, followed by the elimination of whole numbers in each log value, to obtain pure mantissa values, would yield an approximate uniform, flat, and horizontal distribution on the mantissa space of $[0, 1)$, where probability is roughly equal for all mantissa values. Figure 22 depicts the very definition or condition of uniformity of mantissa, regarding the perfect Benford configuration, along with the nine distinct compartments of mantissa corresponding to the nine possible first digits. The height of the "roof" representing the density from the "base" of the M-axis is 1, i.e., $\text{PDF}(m) = 1$, so that the entire area is (1 horizontal length) $\times$ (1 vertical height) $= 1$, as in all statistical distributions. Hence, mini areas — representing digital probabilities for whatever digits and their combinations — are calculated simply and more directly via the reading of the widths on the horizontal M-axis, ignoring the multiplication by 1 (of height). It follows that if a given data set or distribution obeys Benford's Law, then

$$\textbf{Probability}(M_1 < \textbf{Mantissa} < M_2) = M_2 - M_1$$

The uniformity of the mantissa is often referred to as the "general form of Benford's Law".

This is so since the uniformity of the mantissa directly implies the distributions of first-, second-, third-, and all higher-order digits, while $\text{LOG}(1 + 1/d)$ only refers to the distribution of the first-order

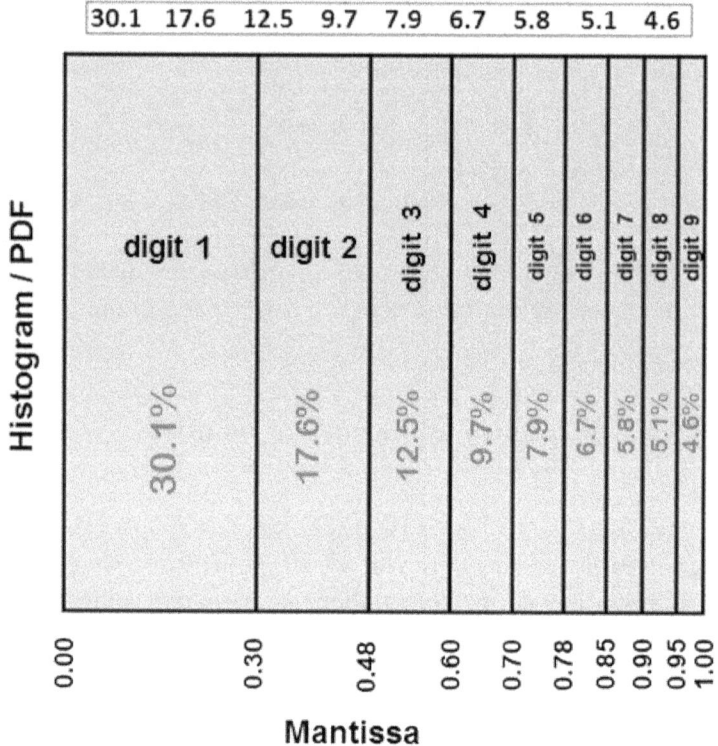

Figure 22.   Nine Mantissa Compartments Corresponding to First Digits

digit. In addition, the general form also implies all the probability dependencies and correlations between the digital orders.

There are nine compartments within the $[0, 1)$ mantissa space, where each compartment points to a unique first digit and whose probability is directly proportional to its width on the M-axis. These nine compartments are: $[0, 0.301)$, $[0.301, 0.477)$, $[0.477, 0.602)$, $[0.602, 0.699)$, $[0.699, 0.778)$, $[0.778, 0.845)$, $[0.845, 0.903)$, $[0.903, 0.954)$, and $[0.954, 1.000)$, as shown in Figure 22. Moreover, each of these compartments is further divided into 10 smaller sections according to the second digit, and each of these second digit sections is further divided into 10 tiny parts according to the third digit. But how did we arrive at these first-digit compartments? What is so special about the numerical edges of the compartments defining them? How were they calculated? The answer is that first digits emerge and switch values exactly at these edges, which are thought of as exponents or powers (i.e., the mantissa), constituting the border lines separating occurrences of the first digits, that is,

$10^{0.000} = 1.00$, $10^{0.301} = 2.00$, $10^{0.477} = 3.00$, $10^{0.602} = 4.00$, $10^{0.699} = 5.00$, $10^{0.778} = 6.00$, $10^{0.845} = 7.00$, $10^{0.903} = 8.00$, $10^{0.954} = 9.00$, $10^{1.000} = 10.00$.

For example, in compartment $[0.301, 0.477)$, mantissa values such as, say, 0.315, 0.348, 0.392, 0.415, 0.458, and 0.473 correspond to the actual numbers 2.065, 2.228, 2.466, 2.600, 2.871, and 2.972, all with the first digit as 2. In other words, the mantissa $[0.301, 0.477)$ corresponds to the actual numbers $[2, 3)$.

The expression for the first digits in Benford's Law could be rewritten as

$LOG(1+1/d) = LOG(d/d+1/d) = LOG((d+1)/d) = LOG(d+1)-LOG(d)$

Probability[First digit is 1] $= LOG(2) - LOG(1) = 0.301 - 0.000 = 0.301$
Probability[First digit is 2] $= LOG(3) - LOG(2) = 0.477 - 0.301 = 0.176$
Probability[First digit is 3] $= LOG(4) - LOG(3) = 0.602 - 0.477 = 0.125$

$LOG(d + 1) - LOG(d)$ is the mantissa area (compartment) representing the probability for the given digit d. The area is actually [the width of the horizontal M-axis] times [the 1 of the vertical height], but more simply it is [the width of the horizontal M-axis], since multiplication by 1 can be ignored.

The natural numbers are $\{1, 2, 3, 4, 5, 6, 7, 8, 9, 10\}$, while the decimal logarithms of these natural numbers are:
$\{\log(1), \log(2), \log(3), \log(4), \log(5), \log(6), \log(7), \log(8), \log(9), \log(10)\}$, that is,
$\{0.000, 0.301, 0.477, 0.602, 0.699, 0.778, 0.845, 0.903, 0.954, 1.000\}$,
and the differences between these logarithms are:
$\{0.301, 0.176, 0.125, 0.097, 0.079, 0.067, 0.058, 0.051, 0.046\}$, being the probabilities for each first digit from 1 to 9.

Chapter 30

Rising or Falling Mantissa Distributions

Figure 23 demonstrates what happens to the first-digit distribution when the log is constantly rising sharply and monotonically. Since here log is defined on (3.0, 4.0) exclusively, log and mantissa are one and the same concept, distributed equally, except for that extra 3 in front of the fractional part. A near-digital equality exists here.

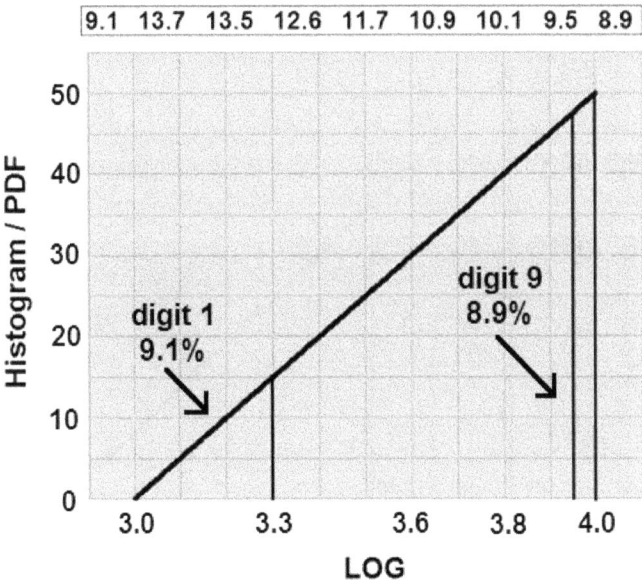

Figure 23. Rising Log Yields Non-Uniform Mantissa and Near-Digital Equality

The area for digit 9, which corresponds to the narrow mantissa subinterval [0.954, 1.000) but of higher density, approximately equals the area for digit 1, which corresponds to the much wider mantissa subinterval [0, 0.301) but of lower density. The fact that the histogram is vertically higher for the higher digits offsets and cancels out their diminishing horizontal widths, resulting in nearly balanced areas for all digits.

Figure 24 demonstrates what happens to the first-digit distribution when the log is only falling sharply and monotonically. Here again, log and mantissa are one and the same concept and are distributed equally, except for the integer 3. An extreme digital skewness exists here in favor of low digits. The area for digit 9, which corresponds to the narrow mantissa subinterval [0.954, 1.000) and very low density, is less than a quarter of 1%, while the area for digit 1, which corresponds to the much wider mantissa subinterval [0, 0.301) and much higher density, is more than 50% of the entire data set. The fact that the histogram is lower for the higher digits, in addition to their diminishing horizontal widths, dooms the high digits, causing them to occupy far smaller areas.

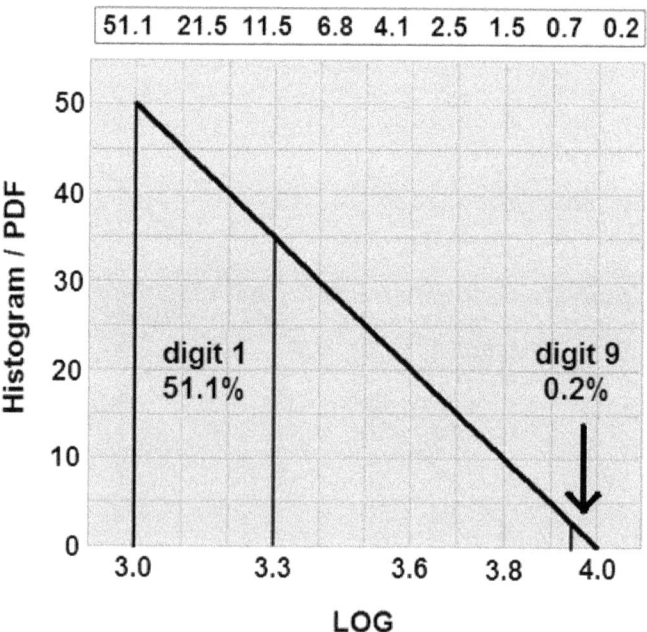

Figure 24. Falling Log Yields Non-Uniform Mantissa and Extreme Skewness

Chapter 31

Uniqueness of k/x Distribution and Its Central Role in Benford's Law

Of all the many known distributions in mathematical statistics, there exists but one unique distribution whose logarithmically transformed values are uniformly distributed, and that particular distribution is k/x. The implication is that Monte Carlo computer simulations of, say, 100,000 values from the k/x distribution, all transformed into their 100,000 logarithmic equivalences, would yield flat and uniform distributions on the log-axis. Mathematical statistics demonstrates that when X is distributed as in $X = 10^{\text{UNIFORM}}$, namely as $X = 10^Y$, with Y uniformly distributed over $[R, S]$, then X is distributed over $[10^R, 10^S]$, having the probability density function $\text{PDF}(x) = [\text{LOG}_{10}(e)/(S - R)]/x$. For example, if Y is uniformly distributed over the interval $(0, 1)$, then X is distributed over $(1, 10)$, and $\text{PDF}(x) = [\text{LOG}_{10}(e)/(1 - 0)]/x = [\text{LOG}_{10}(e)]/x = [\text{LOG}_{10}(2.71828)]/x = 0.434/x$.

One generic rule in mathematical statistics regarding the probability density function of the decimal logarithm of x is that $\text{PDF}[\text{LOG}_{10}(x)] = [x] \times [\text{PDF}(x)]/[\text{LOG}_{10}(e)]$. Applying this rule specifically for the above $\text{PDF}(x) = [\text{LOG}_{10}(e)/(S - R)]/x$ yields $[x] \times [[\text{LOG}_{10}(e)/(S - R)]/x]/[\text{LOG}_{10}(e)]$, or simply $1/(S - R)$, namely a uniform and flat distribution for the decimal logarithm of x.

Since the Benford configuration seeks uniformity of mantissa, and since mantissa is just the fractional part of the logarithm when

$X \geq 1$, it follows that the uniformity of the logarithm of the k/x distribution might endow it with a perfect Benford configuration, assuming some very particular ranges over the x-axis in its definition, so that uniformity of logarithm could translate into uniformity of mantissa. Surely, if the k/x distribution is defined between 10 and 100 for example, then its logarithm is uniform between 1.0 and 2.0, and therefore its mantissa (the fractional part) is necessarily uniform between 0.0 and 1.0. If the k/x distribution is defined between, say, 1 and 10, then its logarithm is uniform between 0.0 and 1.0, and therefore its mantissa (the fractional part) is necessarily uniform between 0.0 and 1.0. But the k/x distribution defined between 10 and 60 is surely not Benford, even though its logarithm is perfectly uniform, and this is so since mantissa here exists only on the shorter and insufficient interval (0.00, 0.78), while having zero height on (0.78, 1.00). The k/x distribution defined between 10 and 200 is also surely not Benford, even though its logarithm is perfectly uniform, and this is so due to its extended range on (100, 200), so that its mantissa is not uniform, having extra height on (0.00, 0.30).

Moreover, the k/x distribution is the only density that perfectly obeys Benford's Law for a range standing between two adjacent IPOT points, such as (1, 10), (10, 100), and (100, 1000), where $(S - R) = 1$. For such ranges, there exists no other distribution that perfectly obeys Benford's Law (with all higher orders considered) except the k/x distribution. On such particular intervals, k/x is unique! Consequently, k/x might be thought of (correctly or mistakenly) as the very definition of the Benford configuration.

Note: The same results and uniqueness are obtained for any range on the x-axis where Max $= 10 \times$ Min, such as k/x over (3.47, 34.7).

Figure 25 depicts the k/x distribution defined over (1, 10). This is the exact mirror image of the uniform logarithm (but depicted as uniform mantissa) over (0, 1) in Figure 22, which shows nine mantissa compartments corresponding to the first digits. Both graphs refer to the same identical distribution, defined over the same range or scope of corresponding values, and this distribution is perfectly Benford with all digital orders considered. Figure 22 refers to its logarithm or mantissa values, while Figure 25 refers to its original raw values.

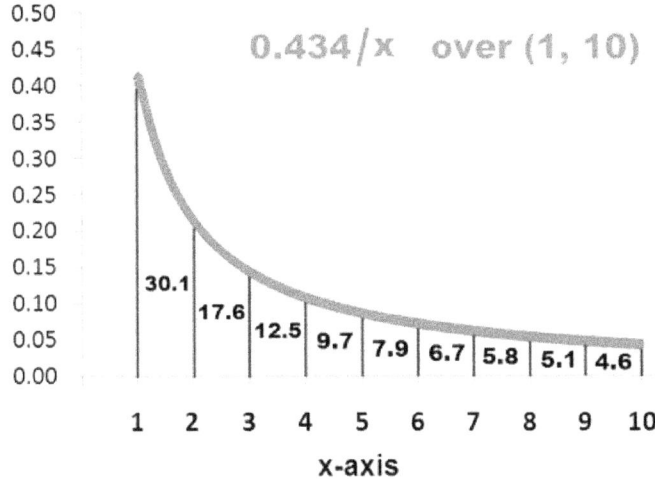

Figure 25. k/x over (1, 10) is the Mirror Image of Uniform Mantissa on (0, 1)

The close relationship between the k/x distribution and Benford's Law of the first digits involving the logarithm of $(1 + 1/d)$, or equivalently $\log(d + 1) - \log(d)$, can be readily demonstrated via the indefinite integral rule in calculus for $x \geq 0$:

$$\int 1/x \ dx = \ln(x) + C$$

For example, in Figure 25, where k/x is defined on (1, 10), the area over the range (1, 2) pertains exclusively to the first digit being 1, and this area is directly evaluated in calculus as the definite integral of k/x over the range of (1, 2), namely $\int (k/x)dx$ from 1 to $2 = k[\int (1/x)dx$ from 1 to 2$] = k[\ln(x)]$ from 1 to 2$] = k[\ln(2) - \ln(1)] = k[\ln(2/1)] = k[\ln(2)] = \text{LOG}_{10}(e) \times [\ln(2)] = \text{LOG}_{10}(e) \times \text{LOG}_e(2)$. Changing the second base from natural e to decimal log via the logarithmic rule $\text{LOG}_\mathbf{A}(x) = \text{LOG}_\mathbf{B}(x)/\text{LOG}_\mathbf{B}(A)$ yields $\text{LOG}_{10}(e) \times [\text{LOG}_{10}(2)/\text{LOG}_{10}(e)] = \text{LOG}_{10}(2) = 0.301$, or 30.1% as predicated by Benford's Law. In the same vein, the area for any other first digit d is easily evaluated via definite integrals as $\text{LOG}_{10}(d + 1) - \text{LOG}_{10}(d)$, namely $\text{LOG}_{10}(1 + 1/d)$. Moreover, the unconditional probability of the second digit being, say, 8, is calculated as the sum of all the areas over the ranges of (1.8, 1.9),

(2.8, 2.9), (3.8, 3.9), (4.8, 4.9), ..., (9.8, 9.9), which all add up as nine distinct definite integrals, totaling 0.0876, or 8.76%, as predicated by Benford's Law. The same correspondence between the Benford unconditional third-digit order and the area evaluations of the k/x distribution is obtained here.

Since the density of the logarithm of the k/x distribution is uniformly and evenly distributed regardless of the range k/x is defined over, its mantissa could also be uniformly distributed and Benford's Law perfectly obeyed — but only if the defined range is carefully calibrated and chosen for that purpose. If the defined range of the k/x distribution is, say, (1, 1000), then each of the three sections (1, 10), (10, 100), and (100, 1000) yields uniformity of mantissa separately in its own right because log distribution is uniform and each section stands exactly between adjacent log integers, rendering the fractional parts (mantissa) uniform locally; therefore, the aggregate or overall mantissa for the entire range (1, 1000) is perfectly uniform as well, and the distribution is perfectly Benford. It should be noted that on the range of (1, 1000), k/x is not uniquely Benford, and there are, in principle, infinitely many other distributions that are perfectly or nearly perfectly Benford as well.

The k/x distribution is also perfectly Benford whenever it is defined between any two points, A the minimum and B the maximum, such that the log difference LOG(B) − LOG(A) is exactly an integer, as in the example of k/x over the interval (1.22835, 12283.5), where the log difference is the integral value of 4. Equivalently stated, k/x is perfectly Benford whenever $B = (10^{\text{INTEGER}}) \times (A)$. From the logarithmic perspective of k/x, the distribution is perfectly Benford whenever it spans an exact integral length on the log-axis from its minimum LOG(A) to its maximum LOG(B), and there is no need whatsoever to align LOG(A) and LOG(B) onto the integers of the log-axis.

Figure 25 directly and visually implies an uneven and skewed second-digit distribution as well. For example, the second digits fully cycle between 1 and 2, as in 1.06, 1.15, 1.29, 1.38, 1.54, 1.75, and 1.97, and they are indeed distributed under a (locally) falling k/x curve of positive skewness between 1 and 2, albeit far less dramatically so in comparison to the (global) first digits fall between 1 and 10. In addition, since "local" skewness of the second digits is visually diminishing as we move from (1, 2) to (2, 3), to (3, 4), and finally to (9, 10), namely as we move from low first digits toward higher first digits, the dependency of the second-order digits on the first-order digits is demonstrated visually and very clearly. Hence, for low first digits, such as on (1, 2), the local curve (of the second order) is relatively skewed, but for high first digits, such as on (9, 10), the local curve (of the second order) is far less skewed and nearly flat. The same line of reasoning and even more refined results are obtained for the third digits, which are yet more equal and by far less skewed, appearing nearly flat everywhere, since the third-digit cycles have a much shorter span on the x-axis.

As it happens, logarithm values of an exponential growth series are also uniformly and evenly spread along the log-axis, albeit in a discrete fashion, as opposed to the continuous way of k/x, so that distances between log values are constant, separated by a fixed value, namely by LOG_{10}(Growth Factor). For example, logarithm values of 25% exponential growth series with a growth factor of 1.25 are uniformly spread along the log-axis, and distances between them are separated by the fixed value of $LOG_{10}(1.25)$. Hence, we note the surprising intimate connection between the k/x distribution and the exponential growth series.

Yet, this connection is actually not really surprising. Exponential growth readings and recording at the end of each period for the evaluation of the growing quantity as time progresses, say in the last minute of each laboratory hour or on the last day of each month, can be algebraically depicted and succinctly expressed as $Q = (\text{InitialQuantity}) \times \text{GrowthFactor}^{\textbf{UniformDiscreteTime}}$, which, in essence, is in the k/x format of $X = 10^{\textbf{UNIFORM}}$, but instead of base 10 to the power of the continuous Uniform as in the case of the k/x, exponential growth is with its unique base of growth factor to the power of discrete Uniform. For example, for an initial quantity of 55 with an 8% exponential growth per period, after the Nth time period, sequential quantities are depicted as $(55) \times 1.08^N$, where the exponent N is the uniformly discrete set $\{1, 2, 3, 4, 5, \ldots, (N-2), (N-1), (N)\}$.

The time variable is discretely selected every minute, month, or year. Surely, time selection must be made fairly, evenly, and uniformly, without focusing on certain time intervals at the expense of other time intervals. By simply choosing to evaluate the growing quantity consistently, say every hour or once a month, fairness and uniformity for the time variable are ensured by default.

Chapter 32

Related Log Conjecture

The most obvious and straightforward way for a distribution to obey Benford's Law is to have its log density itself uniformly distributed on some properly defined and carefully calibrated range (of integral span on the log-axis), so that mantissa in turn could also be uniformly distributed.

But this is not the only way to obtain an approximate or nearly perfect uniformity of mantissa. The alternative to a uniform log density is an upside-down-U curve log density, where the rising part on the left offsets and balances out the falling part on the right, leaving the middle part of approximately flat density as the best representative of the entire curve.

Figure 26 depicts a hypothetical histogram curve of the related logarithm of some random data set. In our terminology here, the log values of the data are said to be *related* to the raw values of the data via the logarithmic transformation $LOG_{10}(\text{data})$. The log curve in Figure 26 starts from the bottom on the log-axis itself and ends all the way down again on the log-axis. On the very left, either

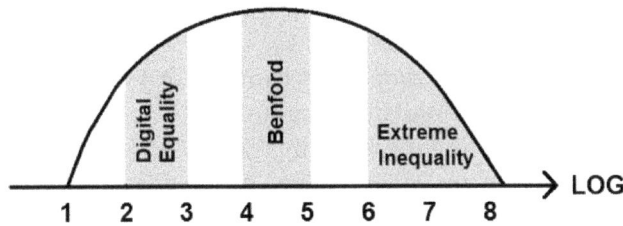

Figure 26. Log Histogram of Random Data and its Digital Variations

high digits are winning slightly or digital equality prevails in the approximate (as in the rising log in Figure 23). Further on toward the center, low digits win only slightly over high digits. Around the center, the Benford configuration is found locally, low digits win strongly, and the graph is approximately uniform and horizontal (as in the flat log in Figure 22). Finally, at the extreme right side, low digits dominate very strongly, much more so than in the usual Benford configuration (as in the falling log in Figure 24). All this suggests some grand trade-off, cancellations, and offsetting effects between the left and right regions, resulting in an aggregate or overall Benford configuration for the entire histogram. Indeed, related log conjecture states that if the distribution comes with a high OOM so that the range of related log histogram is over 3 log units, and especially if it's over 4 log units say, then the data itself is very nearly Benford, assuming that the log curve is smooth without any major irregular bumps, hills, or bulges on the upside and without major cavities, depressions, troughs, or dents on the downside. It is perhaps interesting to note the fact that the log histogram rises on the left, where the slope is generally positive, only to fall on the right, where the slope ends up generally negative, and this observation suggests that if the second derivative is almost always negative, so that the slope is steadily decreasing nearly everywhere, then such smoothness and avoidance of irregularities in the log curve are nearly assured, although minor deviations from this principle are mostly harmless, so that small segments with mild positive values of the second derivative are allowed. Often, disorganized multiple such upward bulges and downward dents tend to cancel each other's adverse effects, leaving the mantissa uniform in the aggregate, unless the bulges and dents are intentionally or accidentally coordinated to occur repeatedly on specific sub-mantissa locations, reinforcing biases for or against specific digits.

A rare exception or counterexample to related log conjecture is depicted in Figure 27, where an unnatural and odd dent is suppressing the log curve around the range of 3.0–3.3 (of digit 1) and elevating the curve around the range of 3.3–3.5 (of digits 2 and 3) so as to rise the highest there. Thus, digit 1 suffers a loss, being reduced from 30.1% to 26.5%, digit 2 gains, being elevated from 17.6% to 23.4%, and digit 3 gains as well, being elevated from 12.5% to 15.0%. Yet, even with this disruptive dent, the overall damage

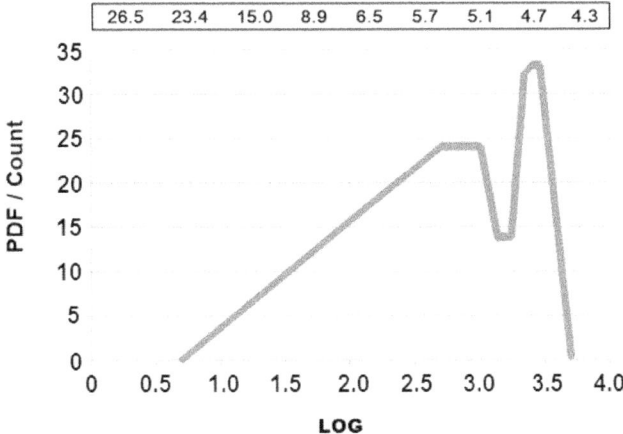

Figure 27. A Rare Counterexample to Related Log Conjecture via a Dent

to the Benford configuration is relatively mild, and the first-digit proportions still decrease monotonically and steadily from 26.5% to 4.3%, while the SSD is 57.1, well below the empirical threshold of 100.

Related log conjecture has in its basis some very profound and persistent geometrical tendencies and forces at play, surpassing any intuition about the enormous possibilities, flexibility, plasticity, and malleability of the mechanizations at work here regarding the overall left–right cancelations and trade-offs. What is remarkable about related log conjecture is that for a relatively smooth curve without any big bulge or dent, almost the same digital result is obtained whenever the entire log curve is displaced to the left or to the right by any amount; in other words, that location on the log-axis does not matter almost, so that shifting the whole curve to the left or to the right by any fractional amount is nearly harmless to Benford behavior. Also, there is no need whatsoever to coordinate the center, plateau, or edges of the log curve along any integral values. In addition, there is no need whatsoever to ensure that the range on the log-axis (namely OOM) spans exactly an integral length either, and non-integral spans, such as 3.4 or 4.7, also yield the Benford configuration. What matters here the most and decisively so is a sufficiently large span on the log-axis, namely a high OOM of over 3, or preferably over 4, so that these left–right cancelations and upsetting effects can take place and be effective.

There is a one-to-one correspondence between (positive) numbers and their logarithms. Two distinct numbers point to two distinct logarithms. And two distinct logarithms point to two distinct numbers. Often, many numerical patterns and essential inner features of the data are revealed by glancing at its log histogram instead of merely staring at the original histogram of the raw numbers. In addition, log histograms often provide superior and more concise visualizations and summaries of data.

The generic shape of the curve in Figure 26 is not some invented fantasy concocted in the excited mind of the young, imaginative, and eccentric statistician, but rather an extremely typical log curve found in almost all random data sets, in Benford data as well as in non-Benford data, financial and accounting data, governmental census data, and scientific and physical data handmade by Mother Nature herself, who strongly favors such aesthetic and round logarithmic curves and almost always dresses herself up with them. She finds straight and linear dresses (of the log of k/x type) to be totally uninspiring, unattractive, and too boring to wear, and she is never able to find a male date for dinner while wearing them. The nature of random and statistical data (Benford or non-Benford), in extreme generality, is characterized by log-gradualism: that related log very rarely starts abruptly high with an initial value; that it very rarely ends abruptly high with a final value; and that it shows a marked curvature which first rises progressively to a certain plateau and then falls gradually down from there, just as humans are first born small and weak, then gradually grow bigger and stronger, achieving their zenith at the age of about 30 or 40, only to grow old and become weak again, even to shrink a bit, and then die. The rare exception to this nearly universal rule is found in the log histogram of exponential growth series, which indeed appears nearly flat and uniform throughout.

The geometrical description of the curve in Figure 26 is termed the **"curvy-closure"** property of the log histogram. The term "curvy" reflects the fact that it curves around, usually gradually and smoothly but not always so, as it could be, on occasions, rough and abruptly changing its direction; but even in such cases, it is almost always being done in an approximately continuous way, so that the curve of the log is without significant breaks, cuts, or gaps, and that it does not have truly disconnected parts. The term "closure" reflects the fact that the curve focuses and warps itself thoroughly and tightly around the log-axis and that it literally "closes" the log-axis, starting from the log-axis itself and terminating there as well, leaving no significant opening or entrance underneath or on its sides. If nearly all real-life data sets are endowed with the curvy-closure property for the histogram of their logarithm, then all they need in order to have the Benford property is simply a sufficient range on the log-axis, namely a high OOM of over approximately 3.

Related log conjecture demands a high OOM in order for all of these cancelations and offsetting effects to be carried out meaningfully and effectively in relation to all nine different fractional compartments of the first digits. For example, there could be no meaningful cancelations whatsoever for the first nine digits occurring under an upside-down-U curve log density of one-unit log width (i.e., an OOM of 1), spanning the narrow log-axis range from 3.0 to 4.0. Such a log curve yields most of its proportions in the bulging middle, around 3.3–3.7 perhaps, pertaining to the first digits 2, 3, and 4, leaving the digits 1, 5, 6, 7, 8, and 9 with lower proportions. As another example, there could be no meaningful cancelations whatsoever for the first nine digits occurring under an upside-down-U log curve spanning the very narrow log-axis range from 3.3–3.6 (i.e., an OOM of 0.3), which pertains perhaps only to the first digits 2 and 3. Data related to such a log curve may not have any of the first digits 1, 4, 5, 6, 7, 8, and 9 whatsoever.

Yet, in the case of an insufficient and narrow range on the log-axis of low OOM, the conjecture could still be applied just as effectively with regard to the second and third digits, which are of much shorter cyclical lengths on the log-axis when compared to the first digits' cyclical length. For example, between 1 and 10 or between 10 and 100, the first digits cycle only once, while the second digits cycle fully 9 times on each of them. Moreover, within each second-digit cycle, such as 4.0 to 5.0 or 70 to 80, the third digits cycle fully 10 times. To recap, between 10 and 100, the first digits cycle only 1 time, the second digits cycle 9 times, and the third digits cycle 90 times. When measured according to log-axis dimensions, the first-digit cycle is of 1-unit log width, the second-digit cycle is of 0.111-unit log width, and the third-digit cycle is of 0.0111-unit log width.

The variable 10 to the power Normal(5.8, 0.22) is a Lognormal-like distribution up to a simple scale adjustment. The log values of the distribution are as in Normal(5.8, 0.22), spanning from around $(5.8) - (3) \times (0.22)$ to around $(5.8) + (3) \times (0.22)$, so that the OOM is only 1.3 approximately, precluding Benford behavior in the first-digit sense, yet the range on the log-axis is wide enough for very meaningful cancelations and offsetting effects in the second- and third-digit sense. To the left of 5.8, log is rising, hence the second and third digits are more equal when compared to their unconditional Benford configuration, while to the right of 5.8, log is falling, hence the second and third digits are more unequal and with greater skewness when compared to their unconditional Benford configuration. Therefore, overall, in the aggregate, the second and third digits are very nearly perfectly Benford. In conclusion, the variable $10^{\text{Normal}(5.8, 0.22)}$ is suffering from severe schizophrenia and confusion, as it feels and acts Benford sometimes, especially with regard to the second and third digits, although at times it is adamantly anti-Benford, especially with regards to the first digits.

The variable 10 to the power Normal(5.8, 0.01) is a Lognormal-like distribution up to a simple scale adjustment. The log values of the distribution are as in Normal(5.8, 0.01), spanning from around $(5.8) - (3) \times (0.01)$ to around $(5.8) + (3) \times (0.01)$, so that the OOM is a tiny value of approximately 0.06, precluding Benford behavior in the first-digit sense as well as in the second-digit sense, since no meaningful cancelations and offsetting effects of the first and second digits exist here, yet the range on the log-axis is wide enough for very meaningful cancelations and offsetting effects in the third-digit sense. To the left of 5.8 log is rising, hence the third digits are even more equal when compared to their unconditional Benford configuration, while to the right of 5.8 log is falling, hence the third digits are a bit more unequal and with greater skewness when compared to their unconditional Benford configuration. Therefore, overall, in the aggregate, the third digits are very nearly Benford, while the first and second digits are decisively non-Benford. Surely, the variable $10^{\text{Normal}(5.8, 0.01)}$ disobeys the law of Benford regarding the first and second digits. On the other hand, it obeys the law when it comes to the third digits.

The fortunate consequence of such short cyclical periods for the higher digital orders is that it is very rare to find data sets or distributions that deviate by much from the Benford theoretical digital distribution of the second order and especially of the third order, unless they are fraudulently concocted. Indeed, the near indestructibility of Benford higher-order digit distributions serves as a splendid tool in data forensics!

Let us demonstrate further the intuition behind related log conjecture via the symmetrical triangular log density, as depicted in Figure 28. Here, the aggregation of all four subintervals standing between the integers 1, 2, 3, 4, and 5 of the log-axis, namely the aggregation of all four local mini-mantissa densities, is clearly horizontal and uniform. The particular view here is the aggregation of mini local mantissa sections, mantissa section by mantissa section. For example, the rising portion on (1, 2) together with the falling portion on (4, 5), being the perfect opposites and complements of each other, are fused and aggregated perfectly into horizontal and uniform mantissa density. The rising portion on (2, 3) together with the falling portion on (3, 4), being the perfect opposites and complements of each other, are also fused and aggregated perfectly into another horizontal and uniform mantissa density. Consequently, these two aggregated horizontal densities in turn are fused into a singular and final aggregated mantissa density which is horizontal and uniform, since its two constituent halves are uniform to begin with. Indeed, the mathematician Lawrence Leemis proved that data arising from the log values of any symmetrical or asymmetrical triangle, namely 10^{Triangle}, where the left edge, the maximum point, and the right edge are all positioned onto integral log values, yield exact Benford behavior.

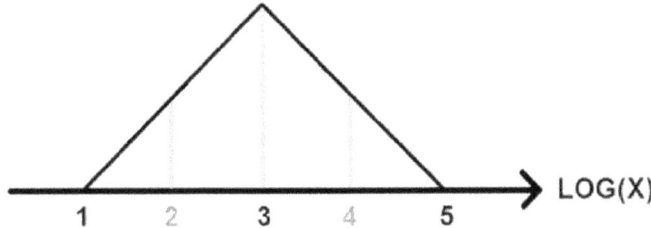

Figure 28. Triangular Log is Perfectly Benford: Mantissa is Uniform

This triangular log density example demonstrates that related log conjecture can be thought of in a generic sense as focusing on the aggregation of all the various local mini-mantissa sections standing between log-integers, weighted by the portion of data points each mini-mantissa section has, so that a mini-mantissa section with lots of points figures in much more heavily than those mini-mantissa sections with only a very few data points. With this view or perspective in mind, related log conjecture states that if OOM is high, then the aggregation of all the various local mini-mantissa sections yields an overall uniform and flat mantissa.

Chapter 33

The Random and Deterministic Flavors in Benford's Law

Not all Benford data sets are created equal, but rather they come with two distinct flavors: the random flavor and the deterministic flavor. By considering the shape of the global log histogram as well as local digital configurations, the distinction between these two crucial flavors in Benford's Law is made without confusing the terms with predictability versus randomness.

The **random flavor** is one where the related log histogram curves around, starting from the log-axis itself, rising, then falling, and finally terminating at the log-axis further to the right, as in the generic format of Figure 26. Locally, such data sets potentially obey Benford's Law only around their center but not on their x-axis locations to the left and right.

The **deterministic flavor** is one where the related log histogram is flat, uniform, and horizontal throughout. Such data sets consistently obey Benford's Law throughout all their parts and subintervals on the x-axis. This is the case of the exponential growth series and the k/x distribution.

Hence, the distinction between these two flavors can be equivalently stated via the global shape of the related log histogram, being flat or curving around, as well as via the way digits behave throughout the entire x-axis range of the raw data locally, on smaller subintervals, around the left, center, and right.

Assuming that a given data set with an x-axis range, say between its minimum of 13 and its maximum of 8729, is very nearly Benford, could we then conclude that parts or sections of the data are also Benford, just because they have been cut out from a whole Benford configuration? For example, is the subset of the data belonging to the subrange between 13 and 652 also Benford? Is the subset of the data belonging to the subrange between 100 and 1000 also Benford? Does the whole endow its Benford property to its parts?

Globally, for the entire data, from the minimum on the left to the maximum on the right, digit proportions are as in $\text{Log}(1 + 1/d)$ overall, as predicated by Benford's Law. Yet, for random-flavored data, local mini digit distributions on smaller subintervals show a remarkably consistent pattern of differentiation, as digits develop from near digital equality on the left for small values, to approximately the Benford digital configuration around the middle, and finally to extreme digital inequality on the far right for big values, where low digits overwhelm high digits and where the digit 1 typically usurps leadership by earning a proportion of 40% or even more than 50% in some cases. This nearly universal pattern in random-flavored data sets is coined **"digital development pattern"**. Coincidentally, and purely by chance, our manipulative political discourse corresponds nicely to the digital development pattern. Our deceptive and fraudulent political narrative is deliberately focused on the "left"–"right" dichotomy, with the "left" supposedly advocating material equality via socialism or oppressive communism and the "right" supposedly advocating unrestrained laissez-faire capitalism, free market economics, and perhaps some harsh material inequality. Then, every other unrelated social issue and public policy questions, shrewdly, artificially, and forcefully get attached to one side or another of this fake binary polarity. The material equality of the political left versus the material inequality of the political right nicely mirrors digital equality on the left of the x-axis versus extreme digital inequality on the far right of the x-axis in data sets.

In order to be able to observe the digital development pattern of the first-digit order, it is absolutely necessary to partition the relevant data range of the x-axis into subintervals standing between integral powers of 10, such as $[1, 10)$, $[10, 100)$, and $[100, 1000)$. This partition is the most natural one in the context of Benford's Law since these points signify the beginnings and ends of all the first-digit cycles. No other partition scheme would do. Other partition schemes not based on integral powers of 10 do not show any development at all and instead confuse the measurement with zigzag or even reversals of skewness.

Orthodox data forensics in the context of data fraud detection via Benford's Law and digit distribution suffer from two pitfalls. The first pitfall is that an increasing number of accountants, economists, businessmen, and other officials are becoming aware of Benford's Law and that, therefore, such sophisticated and well-educated fraudsters might concoct their fake numbers according to the Benford digital proportions so as to avoid detection. The second pitfall is encountered whenever the data itself is of low OOM and therefore not actually Benford in the first-digit sense. This author invented an algorithm, leading to a patent, by applying the digital development pattern to better detect potential fraud. Surely, the well-educated yet mathematically naïve cheater would concoct data without any consideration to development, ignorantly spreading the Benford configuration equally everywhere, on the left, in the center, and on the right, and this can be easily detected. First, a measure called the ES12 of digital skewness relative to the Benford skewness is defined for any given data set or any subset of data, expressing skewness over and above the normal Benford skewed configuration. Second, ES12 is empirically evaluated locally on all AIPOT intervals and recorded for numerous honest and natural real-life data sets, all of which are averaged out together, in order to observe and quantify overall natural and honest development, leading to a standard cutoff or threshold level of development. The true authenticity of the provided data is then decided upon via a comparison of the ES12 development of the provided data itself with the standard threshold of development of these natural and honest data sets.

The measure, termed "excess sum digits 1 and 2" and abbreviated as ES12, is defined as:

ES12 = [observed % of first digits 1 and 2]
    − [Benford % allocation for digits 1 and 2]
ES12 = [observed % of first digits 1 and 2] − [30.1% + 17.6%]
ES12 = [observed % of first digits 1 and 2] − [47.7%]

Figure 29 depicts the digital development pattern for the 2012 global earthquake data set, containing 19,452 data points relating to all earthquake occurrences worldwide during that year and measuring the time in seconds between successive occurrences. Figure 30 depicts the digital development pattern for the data relating to the 2009 government census study on US population centers pertaining to 19,509 cities and towns. These two digital development examples are good representatives of typical real-life random-flavored data sets.

From: To:	1 10	10 100	100 1,000	1,000 10,000	10,000 100,000
1	8.6%	11.3%	15.7%	44.0%	98.6%
2	12.5%	10.2%	14.7%	23.5%	1.4%
3	18.8%	9.8%	13.4%	14.1%	0.0%
4	8.6%	10.2%	11.4%	7.5%	0.0%
5	13.3%	11.0%	10.1%	4.9%	0.0%
6	10.2%	12.6%	9.6%	2.5%	0.0%
7	9.4%	12.1%	9.5%	1.8%	0.0%
8	7.0%	10.2%	8.5%	1.0%	0.0%
9	11.7%	12.7%	7.1%	0.6%	0.0%
Data Points	128	1250	8234	9741	72
% of Data	0.7%	6.4%	42.3%	50.1%	0.4%
ES12	-26.6	-26.3	-17.3	19.9	52.3

Figure 29.   Digital Development Pattern in Global Earthquake Random Data

From: To:	10 100	100 1,000	1,000 10,000	10,000 100,000	100,000 1,000,000
1	5.3%	19.1%	37.3%	46.0%	62.9%
2	8.1%	17.4%	19.7%	20.2%	17.6%
3	7.0%	13.6%	11.6%	10.9%	6.0%
4	9.2%	11.5%	8.6%	6.4%	4.1%
5	11.5%	9.9%	6.3%	5.8%	3.0%
6	13.9%	8.8%	5.3%	3.8%	3.0%
7	13.9%	7.6%	4.3%	2.8%	1.5%
8	17.0%	6.1%	4.0%	2.3%	1.1%
9	14.1%	6.0%	2.9%	1.7%	0.7%
Data Points	1065	8202	7285	2654	267
% of Data	5.5%	42.0%	37.3%	13.6%	1.4%
ES12	-34.4	-11.2	9.3	18.6	32.8

Figure 30.   Digital Development Pattern in US Population Random Data

Both data sets exhibit strong Benford behavior when the entire data set is considered, from the minimum value to the maximum value.

Figure 31 depicts local digital configurations for exponential 5% growth applied for 235 growth periods, starting from 10.0 and ending at 953,861.4 after the 235th period, thus spanning the range of about 10–1,000,000. As can be seen in this table, the Benford configuration is consistently found locally, everywhere along the entire range of data, as well as globally.

| From: | 10 | 100 | 1,000 | 10,000 | 100,000 |
To:	100	1,000	10,000	100,000	1,000,000
1	31.3%	29.8%	29.8%	29.8%	29.8%
2	16.7%	17.0%	17.0%	19.1%	19.1%
3	12.5%	12.8%	12.8%	10.6%	12.8%
4	8.3%	10.6%	10.6%	10.6%	8.5%
5	8.3%	6.4%	8.5%	8.5%	8.5%
6	6.3%	8.5%	6.4%	6.4%	6.4%
7	6.3%	4.3%	6.4%	6.4%	6.4%
8	6.3%	6.4%	4.3%	4.3%	4.3%
9	4.2%	4.3%	4.3%	4.3%	4.3%
Data Points	48	47	47	47	47
% of Data	20.3%	19.9%	19.9%	19.9%	19.9%
ES12	0.2	-0.9	-0.9	1.2	1.2

Figure 31.   Steady Benford without Development in Deterministic Growth

The vast majority of real-life data is of the random flavor (the typical flavor). A tiny minority is of the deterministic flavor (the rare flavor). The choice of the terms "random" and "deterministic" is due to the fact that almost all random real-life data come with such differentiated local digital behavior, while most purely mathematical abstract sequences and series, such as the Fibonacci sequence, exponential growth series, and other such mathematical constructs (associated with exact, deterministic, and predictable values, having no randomness involved), come with very consistent and steady Benford behavior throughout their entire range without any development. An exception to this rule is the random k/x distribution, which is of the "deterministic" flavor! This flavor dichotomy is all about local digital behavior and the shape of the global log, and not at all about predictability versus randomness. Perhaps the terms "differentiated flavor" and "consistent flavor" would make for better terminology, but this would endow the inferior and rare deterministic flavor with the positively sounding adjective "consistent", misguidedly rendering it as if superior to the other flavor with the adjective "differentiated" with its negative connotations of "confused", "inconsistent", or "unstable".

Chapter 34

The Great Prevalence of the Digital Development Pattern in Data

The digital development pattern is extremely prevalent in real-life data, even more so than the Benford phenomenon itself! In other words, practically all random data, Benford as well as non-Benford types (such as those with low OOMs), clearly exhibit a digital development pattern throughout their range. There is almost no exception — it's nearly universal! Moreover, all the theoretical-abstract models in Chapters 12–27, serving as explanations of the Benford phenomenon, come with a decisive digital development pattern. In other words, multiplication processes, partitions, star and planet formations, random growth, data aggregation, and chains of distributions are all decisively of the random flavor, and when local digit configurations coming out of their simulations are examined, development is clearly found. It is delightful to observe the harmony and consistency between real-life empirical data and theory-generated data in the context of development!

In addition to the first-digit development possibility, there also exists a digital development pattern along the line of purely second digits via the partitioning of the x-axis range into subintervals standing between second-digit full cycles, such as 10, 20, 30, 40, or 300, 400, 500, 600. There exists a digital development pattern along the line of purely third digits as well via the partitioning of the x-axis into subintervals standing between third-digit full cycles, such as 10.0, 11.0, 12.0, 13.0 or 70.0, 71.0, 72.0, 73.0. The digital development patterns along the second and third digital lines are even more decisive and prominent than the development along the first digital line, since this can be found unfailingly and reliably even in data with very low OOM! Hence, the developments of the second and third digits serve as splendid tools in data forensics!

Within each of these narrower subintervals of the x-axis — partitioned according to the occurrences of the cycles of second or third digits — we define a measure of second or third digital skewness relative to the Benford skewness, which expresses skewness over and above the normal unconditional Benford second or third skewed configuration.

The measure for second digits is termed "excess sum digits 0 to 4" and abbreviated as ES04:

ES04 = [observed % of second digits {0, 1, 2, 3, 4}]
       − [Benford % allocation for second digits 0 to 4]
ES04 = [observed % of second digits {0, 1, 2, 3, 4}] − [54.7%]

The measure for third digits is termed "third excess sum digits 0 to 4" and abbreviated as TES04:

TES04 = [observed % of third digits {0, 1, 2, 3, 4}]
        − [Benford % allocation for third digits 0 to 4]
TES04 = [observed % of third digits {0, 1, 2, 3, 4}] − [50.49%]

Data with a low OOM cannot be forensically analyzed via Benford's Law in the first-order sense since they do not obey the law *a priori* and can only be checked perhaps against Benford second and third orders. Yet, since cheaters often tend to concoct numbers quite randomly in their heads and perhaps uniformly so, and since the second and third orders are themselves nearly uniform and even, the forensics here may be ineffective or even useless. Luckily for the data forensics specialist, the data could still come under the closer scrutiny of the digital development pattern along the line of purely second digits or along the line of purely third digits, both of which require only a modest OOM for the data in order for the forensics to be performed effectively.

Caution should be exercised when the second-digit development ES04 is examined on the basis of purely second digits, such as on (10, 20), (20, 30), (30, 40), and so on up to (90, 100), so as not to confuse it with the intrinsic second-order digital fluctuations due to dependencies of the second digits on the first digits. As mentioned in Chapter 3 regarding the dependencies and correlations between the orders, second digits exhibit more skewness than their base unconditional skewed configuration on say (10, 20), where the first digit is 1, and they are more equal on (90, 100), where the first digit is 9. In other words, due to dependency, there exists more skewness on the left and more equality on the right between IPOT values — the polar opposite of development with more equality on the left and more skewness on the right overall from the minimum to the maximum values. Clearly, dependency and development push in opposite directions. The net fluctuation in second-digit configurations relative to their base unconditional proportions is governed by the confluence of the digital development pattern together with the order dependency pattern. These two patterns exist side by side and compete fiercely for dominance. When OOM is high, say 3 or 4, these two distinct patterns mix and fuse together in relative harmony, each flaunting its power and influence, and each pattern can be seen clearly, as the dependency pattern is deciphered cyclically between 1, 10, 100, 1000, and such, while the development pattern is deciphered only globally, in the aggregate, from the minimum to the maximum values. When OOM is low, say 1, that upside-down-U log curvature of the development pattern is very sharp, dramatic, and abrupt, as the entire development is compressed onto a singular AIPOT interval, and so the development pattern easily overwhelms the dependency pattern, nearly obscuring it completely.

Note: The second digital development pattern ES04 could also be detected along the basis of the first digits on subintervals standing between AIPOT such as (1, 10), (10, 100), (100, 1000), assuming a sufficiently high OOM, namely at least two such AIPOT intervals for comparison, but preferably three for better comparison. The second digits actually run nine full cycles on each of these long AIPOT intervals, and therefore the widely aggregated second development can be seen even more decisively. As an example, we calculate the second digits on, say, (1, 10), and find that they are a bit more equal than the second digits on (10, 100), which are of greater skewness. Then, we calculate the second digits on

(100, 1000) and find that they are of even greater skewness when compared to those on (10, 100). The advantage of examining the second digital development pattern along the basis of the first digits is that there is no interference or potential confusion due to dependency (i.e., the correlations between the first and second orders) because, here, whole AIPOT intervals are examined, each containing the entire repertoire of the first digits. In addition, the third-digit development pattern TES04 could also be easily detected along the first-digit basis of AIPOT intervals, such as (1, 10), (10, 100), and (100, 1000), assuming at least two or three such AIPOT intervals for a good comparison. The third-digit development pattern could also be easily detected along the second-digit basis, such as on (1, 2), (2, 3), and (3, 4), given a sufficient span on the x-axis.

Chapter 35

The Absence of the Digital Development Pattern in k/x Distribution

Certainly, the case of k/x distribution is quite central in the field of Benford's Law, yet its highly consistent Benfordian feature throughout its range, on its left, center, and right, totally lacking a digital development pattern, renders it quite odd, very rare, and totally irrelevant to almost all real-life random data. "Paradoxically", k/x is the only distribution that can perfectly obey Benford's Law (with all higher digital orders considered) for a short range standing between two adjacent integral powers of ten points, such as (1, 10) or (10, 100), and this unique feature renders the case of k/x distribution quite exceptional in the field of Benford's Law.

Such is the seductive power of the k/x distribution in the context of Benford's Law that some misguided authors and overly enthusiastic students of Benford's Law begin their article or essay by basing it on some assumption or feature regarding the k/x distribution and then proceed to draw far-reaching conclusions, mistakenly extrapolating the odd case of k/x to all real-life random data. Such a regrettable trend has led to several erroneous conclusions, published in respectable journals and officially certified by expert mathematicians as true. This author has taken on the dissenting role of an agitator as well as a prophet of doom, preaching the virtue of separating the random from the deterministic and of becoming aware of this crucial distinction in the field, and predicting the encountering of contradictions between the empirical and the theoretical in all such misguided pseudo-mathematical endeavors.

Chapter 36

Benford's Law in Its Purest Form

The simple, intuitive, and straightforward case of the k/x distribution explains and absolves Mother Nature of the accusations that she is actually too compulsive, free-spirited, and very much down-to-earth to be able to carefully adjust her quantities in such a complicated way so that the fractional parts of the logarithms are evenly and uniformly distributed. It can be argued that the main feature of Benford's Law is to provide Mother Nature with a very simple and natural model or template of the positively skewed k/x distribution, with its tail falling to the right, where the frequency of quantitative occurrences (the density height at x) is inversely proportional to quantity ($\propto 1/x$), and to ask from her to behave in such a way that, in the aggregate, overall, from the minimum on the left to the maximum on the right, in between them in the center, and everywhere, when all considered, her quantities are in the spirit of the k/x distribution. The fact that the fractional parts of the logarithms are evenly and uniformly distributed is incidental, secondary, and simply a minor mathematical consequence, and Mother Nature is not even aware of this fact. She has never even learned of the strange concept of the logarithm in her youth at her school and would certainly not have been able to understand her teacher if taught the subject.

We regard the concept of the logarithm as too abstract and too subtle to be involved in the expressions of the primary or primordial laws of nature, although it is acceptable and very much expected to appear in many of the secondary results, which may be succinctly expressed logarithmically. For example, Newton's law of universal gravitation, $F_G = G \times M_1 \times M_2/R^2$, is a straightforward result expressing the fact that $M_1 \times M_2$ serves as an accounting scheme designed to sum up at the macro level all the gravitational forces occurring at the micro level between the two masses consisting of numerous tiny elemental parts (say protons and neutrons), where the total number of these tiny parts for a given object expresses the value of its mass. Each tiny elemental part within Object 1 attracts each tiny elemental part within Object 2. The net macro resultant gravitational force between these two macro objects is the totality of these micro interactions and which is accounted for by the multiplication of the number of these elemental parts in Object 1, namely M_1, by the number of these elemental parts in Object 2, namely M_2; therefore, $M_1 \times M_2$ expresses the influence of the masses on the force.

The $1/R^2$ part is simply the result of three-dimensional geometry expressing the fading density of gravity distributed evenly over the surface of an ever-expanding sphere and emanating from a piece of mass at its center. Since the area of the surface of a sphere is $4\pi R^2$, it follows that $F_G \propto 1/R^2$ expresses the gravitational force per surface area. It would be very strange, or rather a bit unbelievable, had Newton offered us instead an expression involving logarithms, such as $F_G = G \times \mathrm{LOG}(M_1 \times M_2)/R^2$, or $F_G = G \times M_1 \times M_2/\mathrm{LOG}(R^2)$, although we would be forced to accept it if verified experimentally, suppressing our dislike and our intuition against it. Surely, the explicit use of the logarithm famously appears in Boltzmann's entropy formula in thermodynamics, being expressed as $S = k_B \times \mathrm{LOG}_e(W)$. Yet, this is considered a mere result and consequence of the more primordial and basic laws of nature, together with complex calculations in physics and mathematical statistics, known as statistical mechanics.

Hence, while avoiding the usage of logarithm, we shall state that Benford's Law does not directly address the issue of *digital distributions* at all, nor does it address the format of the *logarithmic curve* in any direct way; the essence of the law is really the simple and natural statement that *quantities* in the world are distributed in the aggregate, as in the generic k/x distribution. The focus on quantities renders it universal, independent of any number system, and independent of any base B. The part in Benford's Law of the $\mathrm{LOG}(1 + 1/d)$ statement involving the odd concept of the logarithm is merely a mathematical consequence regarding the relative occurrences of the first digits whenever we superficially dress up quantities in the coat (digits) of our invented positional number system. Benford's Law, in essence, states: that falls in the histogram, such as in k/x^2, k/x^3, and k/x^4 distributions, are too sharp and too steep; that falls in the histogram, such as in $k/x^{1/2}$, $k/x^{1/3}$, and $k/x^{1/4}$ distributions, are too mild and too leveled; and that the true generic rate of fall in quantities found in nature is of the k/x^1 type. Random-flavored data is never as the k/x distribution locally, and log is almost never flat and uniform locally, since the logarithmic curve rises on the left and falls on the right, and local digit configuration fluctuates as predicated by the digital development pattern. However, globally, from the minimum to the maximum, the log curve averages out (on the mantissa level between log integers) to be uniform, as in the k/x distribution, and overall the digit distributions are Benford for the entirety of the data.

Moreover, not only does the k/x distribution provide Mother Nature with an ideal model and inspiration and not only does it implore her to fall to the right and be skewed in such a deliberate manner in the aggregate, neither too sharply nor too mildly, but also this is so in a more concrete and quite tangible way whenever all quantities are being forced to express themselves exclusively on any one particular AIPOT interval, as for example on the (1, 10) interval. This is accomplished by adjusting and moving the decimal point either to the right or to the left so that all quantities fall on the (1, 10) interval. For example, quantity 1358 would be expressed as 1.358, and quantity 0.0962 would be expressed as 9.62, while quantity 7.03 would be left as such without adjusting the decimal point. Histograms of any type of real-life Benford data set adjusted to fall on any one particular AIPOT interval, such as the (1, 10) interval, are indeed very nearly as in the k/x distribution!

As for one real-life concrete data example, the 2012 earthquake data set regarding the time between 19,452 consecutive occurrences, forced to fall on (1, 10) via such decimal shifts, is depicted in Figure 32 as the grey histogram with nine bins, standing between the integers 1–10 and being superimposed with the black line representing the theoretical $8447/x$ distribution evaluated at the midpoint of each bin, namely at 1.5, 2.5, 3.5, and so on up to 9.5. The choice of 8447 was obtained by forcefully endowing the k/x

distribution with an area of 19,451 instead of the usual area of 1, and this was done by multiplying $\text{LOG}_{10}(e)/x$, or $0.4343/x$, by the factor 19,451, yielding $(19451) \times (0.4343)/x$, that is, $8447/x$. In order to get the histogram and the k/x curve on an equal footing for proper comparison, we equate the entire k/x area to the entire area of the earthquake, which is 19,451, since the earthquake histogram is constructed between the integers, thus having a bin width of 1, so that the area of each bin is simply its frequency (i.e., its height), and consequently the sum of the areas of all the bins of the earthquake data is simply the total number of data points, which is 19,451. Note that 19,452 earthquake occurrences yield 19,451 time intervals between them.

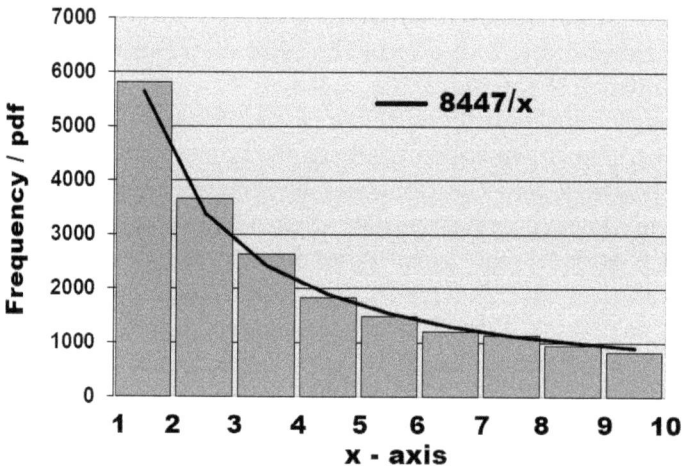

Figure 32.   Good Fit for Earthquake Data Expressed on (1, 10) and $8447/x$

Astute readers might suspect this author of tricking them by dressing up the uniformity of the mantissa in the spirit of the k/x distribution, since adjusting and moving the decimal point so as to force the numbers to fall on (1, 10) is equivalent to deleting or ignoring the 10^{INTEGER} part in the expression $X = 10^{\text{Log}(X)} = 10^{(\text{INTEGER}+\text{MANTISSA})} = (10^{\text{INTEGER}}) \times (10^{\text{MANTISSA}})$. These highly perceptive readers feel cheated because the earthquake histogram part in Figure 32 is indeed (also) 10 to the power of the mantissa of the earthquake numbers. In other words, the histogram is the fusion of all the earthquake numbers combined together according to the IPOT location, which (they claim) is in essence just a scheme based on the cycles of the mantissa.

This author denies the accusations of deception leveled against him but admits that the earthquake histogram in Figure 32 is indeed equivalent to the 10^{MANTISSA} of the earthquake numbers, then further excuses himself by deferring to the fact that the mantissa of the k/x distribution defined over $(1, 10)$ and the mantissa of the Benford earthquake data (or the mantissa of any Benford data) are both indeed uniform. The judge in his trial at the "supreme honesty court", while cross-examining the author, harshly reprimanded him by asserting, "It is apparent that by dividing the earthquake data according to mantissa sections standing on the edges of integral logarithm values, such as 0, 1, 2, and 3, you have employed both the concept of the logarithm and the concept of the mantissa!" The defendant author then retorted, "Your Honor, let me explain. I was not paying the slightest attention to mantissa or logarithm; it's just that I have heard some vague claims and rumors that the first digits of almost any real-life data occur with some very particular pattern, which is called Benford's Law, so it aroused my curiosity, but the strangest part of it was that they claimed that nearly every real-life data set resembles the k/x distribution in some way; so in order to verify this, I was merely dividing and grouping the earthquake data according to the cycles of the first digits, which occur exactly between 1, 10, 100, 1000, and so forth. Then, I attempted to adjust the decimal points so that all quantities from all these groups would fall on the same $(1, 10)$ interval, simply as a way to condense and aggregate all first-digit manifestations occurring anywhere. I now rest my case." Whether history will absolve the author of deception or convict him of it is not yet known.

Chapter 37

Constant Base Raised
to a Random Power

The fundamental logarithmic relationship $X = 10^{\text{Log}(X)}$ simply restates the definition of the logarithm: "Log x is the number to which 10 must be raised in order to get the number x".

This facilitates the following summary of two distinct mathematical formats, one for the deterministic flavor and another for the random flavor in Benford's Law:

$$10^{\text{Uniform(a, b)}}$$
$$10^{\text{Normal(m, sd)}}$$

The Uniform case is the one where the logarithm is Uniform, thus pertaining to the k/x distribution and the deterministic flavor. The Normal case is the one where the logarithm is Normal, thus pertaining to the Lognormal-like distribution and the random flavor, except for a minor adjustment via the multiplicative factor $\text{LOG}_{10}(e)$, since the Lognormal is defined as e raised to Normal power, not as 10 raised to Normal power.

In order to ensure Benford behavior, it is necessary to carefully calibrate the parameters a, b, m, and sd.

For the case of Uniform powers, we seek an exact integral range on the log-axis, or at least approximately so, via the calibration of $(b - a) = \text{Integer}$.

For the case of Normal powers, we may first wish to avoid negative values by assigning a positive m value to be at least three times bigger than the sd value, perhaps for simplicity sake, but this is not really necessary. What is really necessary here for Benford behavior is a wide range on the log-axis via a large *sd* value, namely an OOM bigger than 3 for resultant data. The three sigma rule in empirical statistics states that 99.7% of the data occurs within three standard deviations of the mean in Normal distributions; therefore, in the approximate, the parameters *m* and *sd* should be chosen so that $OOM \approx [m + (3) \times (sd)] - [m - (3) \times (sd)] = (6) \times (sd) > 3$, i.e., so that *sd* must be at least 0.50.

$$B^{Uniform(a, b)}$$
$$B^{Normal(m, sd)}$$

If the base is non-decimal, i.e., $B \neq 10$, the above constraints on the parameters are adjusted slightly. Our measuring rod and benchmark is still the decimal base-10 number system, as digits, quantities, numbers, and logarithms are all based on our standard decimal base, which is also the basis of all our empirical digit distributions and practical compliance tests regarding Benford's Law.

For the case of Uniform powers, the general prerequisite of exact Benford behavior for any non-decimal base B is restated, requiring $OOM = LOG_{10}(B)(b - a)$ to be exactly an integer, or at least approximately so, which guarantees Benford behavior.

For the case of Normal powers, a relatively big base, i.e., $B > 10$, would allow for a smaller *sd* value, since the big size of the base naturally yields an already larger dispersion and more variability in the resultant quantities. A relatively small base, i.e., $B < 10$, would require a bigger *sd* value in order to arrive at the Benford configuration, since the small size of the base yields less dispersion and less variability in the resultant quantities. Approximately, the parameters *m* and *sd* should be chosen so that $OOM \approx LOG_{10}(B) \times [[m + (3) \times (sd)] - [m - (3) \times (sd)]] = LOG_{10}(B) \times [(6) \times (sd)] > 3$, so that $LOG_{10}(B) \times (sd) > 0.50$, i.e., so that $sd > 0.50/LOG_{10}(B)$ and, more concisely, so that *sd* must be at least $0.50 \times LOG_B(10)$.

As one concrete example, $B^{Normal(3, 0.5)}$ with a fixed mean value of 3, a fixed *sd* value of 0.5, and employing generic and variable base *B* values yields digital configurations which converge to Benford as the chosen value for the base B increases. Figure 33 depicts some possibilities of these types of distributions, showing their resultant OOM and SSD values of the first digits for a variety of base *B* values in {2, *e*, 4, 5, 7, 11}, obtaining 10,000 simulations for each *B* case.

DISTRIBUTION	SSD	OOM
$2^{\text{Normal}(3,\,0.5)}$	701.4	0.9
$e^{\text{Normal}(3,\,0.5)}$	390.6	1.3
$4^{\text{Normal}(3,\,0.5)}$	68.5	1.8
$5^{\text{Normal}(3,\,0.5)}$	33.1	2.1
$7^{\text{Normal}(3,\,0.5)}$	3.2	2.5
$11^{\text{Normal}(3,\,0.5)}$	0.8	2.8

Figure 33. Large Base is Necessary for Benford Behavior

Interestingly, and consistently with earlier discussions, only after the OOM reaches approximately the value of 3 does the SSD finally fall below 2, and the distribution is nearly perfectly Benford.

$B^{\text{AnyRisingThenFallingDistribution}}$ pertains to the generic random flavor, which is indeed Benford as per the related log conjecture, assuming a large enough OOM of the entire resultant random process and assuming some minimal smoothness, continuity, and consistency in how the related log histogram rises and falls, without any gaps, large dents, or bulges, as well as a beginning and an end near the log-axis itself (i.e., having the curvy-closure property).

SECTION IV
General Results

Chapter 38

General Results in Benford's Law

Code and index numbers, lottery numbers, phone numbers, social security numbers, driver's license numbers, passport numbers, assigned ID numbers, post office zip code numbers, and such do not conform to Benford's Law in any way, and their digital distributions are uniform with equal proportions for each digit. For these types of numbers, each digit is randomly selected with equal probability independently of its adjacent digits, so that digits occur with an equal 10% probability. Put another way, a lottery "number" of six digits is, in reality, not a single number as is commonly perceived, but rather a set of six totally independent and unconnected digits, necessitating six different decisions. The common denominator in all of these types of "numbers" is that they (or any single digit within them) do not represent quantities or counts of anything in the physical world. For example, if Frank's phone number is 2474633 while that of Simon is 9884589, then it would be ludicrous or absurd to conclude that Simon (compared to Frank) is older, heavier, wiser, taller, richer, or anything else quantitatively, since these digits do not stand for any quantities out there in the physical world.

Artificial numbers that are influenced by human thoughts, such as ATM withdrawal numbers or a particular price schedule that is made less to reflect cost and more to attract clients and increase sales, do not obey Benford's Law. Quite often, prices are set to fall below a certain psychological threshold, such as $4.99, which is perceived by clients as much lower than $5.00. Typical amounts taken from an ATM machine, such as $20, $240, $100, $60, and $80, are too even and very particular, invented in the minds of the account holders.

Another important exception to Benford's Law is the case of numbers with a built-in maximum or minimum value, either artificially with human intent or due to some particular limits on values.

The set of all **physical constants**, 337 in total, assuming the metric system and other scales and units, can be found at http://physics.nist.gov/cuu/Constants/Table/allascii.txt. The list includes items such as Planck and Boltzmann constants, Bohr radius, constants relating to gravitational, electromagnetic, and quantum forces, as well as others in physics and chemistry. Remarkably, its first-digit distribution is {34.7, 19.0, 8.9, 8.3, 8.3, 7.1, 3.3, 5.1, 5.3}, which is quite close to Benford in spite of the scarcity of values here, that is, its small data size. This interesting result cannot be explained as simply another manifestation of the Benford phenomenon in the natural world due to multiplication processes, partition processes, data aggregations, chains of distributions, and so forth, since the set is not at all about any single physical issue or phenomenon; rather, this unique set is about the format and regularities governing all known phenomena in nature. It surely lends Benford's Law some more weight and even mystique. Significantly, calculating the first digits of physical constants assuming other scales and different units do not change results by much for the most part, as in the scale invariance principle.

Exponential growth series are expressed as $\{B, BF, BF^2, BF^3, \ldots, BF^N\}$, where B is the initial base value, N is the number of growth periods, $P\%$ is the constant percent growth rate, and F is the constant multiplicative factor related to the growth rate, as in $F = (1 + P/100)$. The consideration here is of all the quantities during all the time periods as a singular data set. The vast majority of exponential growth series are nearly perfectly Benford, and only a tiny minority are non-Benford. In extreme generality, the larger the number of growth periods (yielding large data sets), the closer the growth series to Benford, and typically (assuming that the growth rate is sufficiently high), just 100 or perhaps 500 periods are enough, depending on the particular growth rate chosen and the desired level of accuracy in terms of obtaining a good fit to Benford. For a **low** growth rate, it is necessary to have the OOM near an integral value so that the span of its flat and uniform logarithm values is approximately an integral range on the log-axis, namely that the exponent difference LOG(Last) − LOG(First) should be as close as possible to an integer, so that mantissa is uniform. For a **high** growth rate and considering plenty of growth periods, this integral constraint is superfluous and can be ignored, since the series frequently passes IPOT values every few periods, thus lying over numerous AIPOT intervals, rendering any partial or incomplete (extra) AIPOT interval insignificant in the grand scheme of things. In other words, that extra interval constitutes a very small disruptive percentage when compared to the entire series.

Besides possibly an integral OOM being the first prerequisite, the second general requirement regarding Benford behavior for any growth series is that $\text{Log}(1 + P/100)$ must not equal a rational number, or even approximately so, i.e., $\text{Log}(F) \neq L/T$, where L and T are integers. Other neighboring $P\%$ growth rates near such $P\%$ growth with a rational equality for its $\text{Log}(F)$ also experience significant deviation from the Benford configuration, with the degree of deviation depending on the distance from it, unless they are of an infinite length or, in practical terms, unless they are with an exceedingly large finite number of periods so as to overcome the proximity to the problematic growth.

For example, exponential growth series near 151.2% deviate significantly from Benford and are called rebellious or anomalous, since $\text{Log}(1+151.2/100) = \text{Log}(1+1.512) = \text{Log}(2.512) = 0.4 = 2/5$.

In another example, exponential growth series near 58.5% deviate significantly from Benford and are called rebellious or anomalous, since $\text{Log}(1 + 58.5/100) = \text{Log}(1 + 0.585) = \text{Log}(1.585) = 0.2 = 1/5$.

There exist 774 rebellious zones from 1% to 900%, but many of them are mild, with SSDs not exceeding approximately 20; only 56 of these zones are significant and truly disruptive, with SSDs over 100.

Standard deterministic exponential growth series with fixed parameters, involving the consideration of all the quantities during all the time periods as a singular data set, has the steady local Benfordian digit configuration throughout its entire range without development, being of the deterministic flavor. On the other hand, Ross models regarding the collection of the last term $\{BF^N\}$ of numerous and distinct exponential growth series of random parameters all come with a decisive digital development pattern, indicating that these models are unquestionably of the random flavor and not of the deterministic flavor.

Prime numbers are decisively non-Benford. Carl Friedrich Gauss discovered the unexpected and surprising connection between primes and logarithms, expressing the prime number density as $1/\ln(N)$, which implies a mild and gentle fall in the histogram to the right. This result is known as the prime number theorem. Hence, low first digits are a bit more prevalent than high first digits for primes up to approximately 10,000, and then they rapidly approach digital equality in the limit toward infinity:

Primes between 10 and 100: $\{19.0, 9.5, 9.5, 14.3, 9.5, 9.5, 14.3, 9.5, 4.8\}$

Primes between 100 and 1000: $\{14.7, 11.2, 11.2, 11.9, 9.8, 11.2, 9.8, 10.5, 9.8\}$

Primes between 1000 and 10000: $\{12.7, 12.0, 11.3, 11.2, 10.7, 11.0, 10.1, 10.4, 10.6\}$

The **Balls and Boxes** model is a scheme of the random throwing of L distinguishable balls into N stationary and distinguishable boxes, where each ball is equally likely to fall into any box, regardless of the number of balls already inside the boxes during each throw. This yields a positively skewed distribution of the balls, but not the Benford configuration. An alternative model involves declaring L undistinguishable balls as already residing inside N distinguishable boxes and postulating that all possible box configurations are equally likely. A box configuration is the entire specification of the states of all the boxes in terms of how many balls each particular box contains. The term "configurational entropy" refers to the assumption that all possible system configurations are equally likely. This last model also yields a positively skewed distribution of the balls, but not the Benford configuration. Surely, these two distinct descriptions of randomness or arrangements of balls inside boxes do not yield identical results, either quantitatively or digitally. The physicist Oded Kafri has presented the latter model of the balls and boxes idea in order to discover a mathematical connection between the principles of entropy in physics and Benford's Law for some particular situations where particles are randomly spread throughout certain systems according to the laws of thermodynamics. The physicist Don S. Lemons had worked on another model in order to discover a mathematical connection between the principles of entropy in physics and Benford's Law.

The **Fibonacci series** {0, 1, 1, 2, 3, 5, 8, 13, 21, 34, 55, 89, 144, 233, 144 + 233, ... } begins with {0, 1} as two arbitrarily chosen starting points. Subsequent elements beginning with the third are the addition of the previous two elements, expressed algebraically as $X_N = X_{N-1} + X_{N-2}$. Interestingly, even though the series is defined in terms of additions, it approaches a repeated multiplication process very early on, with the golden ratio 1.61803399 as the factor, since successive elements can be approximately obtained by simply multiplying the previous element by the golden ratio. This explains why the Fibonacci series is almost perfectly Benford. For example, the 12th element, namely 89, is approximately equal to the golden ratio times the 11th element, 55. That is, $(55) \times (1.61803399) = 88.99 \approx 89$. Algebraically, the golden ratio approximation states that after the first few elements (say from the 11th element on), the series becomes nearly a standard exponential growth series expressed as $X_N = (1.618) \times X_{N-1}$.

It is also noted that due to its relatively high growth rate of 61.8%, the Fibonacci series passes through an IPOT number once every five terms approximately, thus the requirement of an integral order of magnitude for Benford behavior can be easily ignored, and we need not worry about where we start and where we end, only that we must consider plenty of elements. As a check, $1.618^5 \approx 11 > 10$, so that in about five terms, the cumulative increase is over tenfold, and an IPOT number is passed.

The **mixture of distributions**, namely the aggregation of a large number of related as well as totally disparate and unrelated distributions, each defined over the positive x-axis, obeys Benford's Law in the limit as the number of distributions goes to infinity. This is also called "the distribution of all distributions".

The mathematical model, however, cannot be a valid explanation for the nearly perfect Benford behavior of data sets regarding single-issue (single-source, non-mixed) physical and scientific phenomena, such as the time between earthquake occurrences, river flow, population count, pulsar rotation frequency, and the half-life of radioactive material. The mixture of distributions model does not show any immediate or obvious relation to the Benford way Mother Nature generates her physical quantities, topic by topic, issue by issue, and phenomenon by phenomenon. It is very hard, or rather impossible, to argue that river flow, earthquake timing, pulsar rotation, population count, and so forth are each the results of some aggregation of numerous invisible, mysterious, and unrelated mini distributions.

This author has performed a time-consuming and tedious data experiment related to this theory for several days, collecting and mixing real-life Internet data from a wide variety of related and totally unrelated sources or websites, resulting in a large set of 34,269 positive numbers. The results resolutely confirmed the theoretical-mathematical expectation, with the first-digit distribution of the entire set of numbers calculated as {28.8, 16.4, 12.4, 9.8, 8.3, 7.3, 6.1, 5.7, 5.3}, with the SSD being a low value of 4.6, indicating strong compliance with Benford's Law. Yet, conceptually, putting all these in context, this mixture of mostly unrelated numbers does not represent seconds, hours, years, kilograms, pounds, dollars, euros, miles, feet, meters, hertz, solar mass, or any other unit or quantity, nor does it convey any specific data-related message. This data set does not represent any particular physical entity or physical concept, nor does it measure under any single scale or unique unit. Mathematically, though, the digits of such data sets are demonstrated to be perfectly Benford in the limit, yet this explains nothing but itself. Detailed discussions about this data experiment, the mathematician who provided the proof on the mixture of distributions, and its harmonious connection to the related log conjecture can be found in Chapters 55, 69, and 110, respectively, of the author's 2014 book on Benford.

The **Exponential distribution** is never perfectly Benford, regardless of the value assigned to the parameter lambda (λ), but it is fairly close to it. Fascinating repeated patterns are seen whenever the nine first-digit proportions are (vertically) plotted versus the independent parameter λ (horizontally). In other words, under the consideration of the first digits as the dependent variable being a function of the independent variable λ parameter, we draw nine such functional curves for all the nine digits. Each individual digit nicely oscillates consistently and devotedly around its central value of $\text{LOG}(1 + 1/d)$ indefinitely, above and below it, in ever-widening cycles as the parameter λ increases. Unfortunately, the nine cycles of the nine digits are not synchronized; rather, they are continuously out of step with each other, making it completely impossible to find one particular value of the parameter λ that is able to simultaneously seize all nine digits resting exactly at the Benfordian centers of their cycles. On the other hand, a chain of two distributions in the form of Exponential(Uniform(0, M)), being the collection or aggregation of diverse Exponential cases, where the parameter λ takes on various values within (0, M), leads to a near-perfect Benford configuration. The beneficial role that the chain plays here is its function in smoothing out all the oscillations and deviations of all the nine digits from their respective Benfordian centers. The net effect of the act of chaining here is to average out distinct digital configurations and thus to cancel much of the deviations, leading to the Benford configuration.

Limits on the value of the base B were suggested by the applied mathematician Frank Benford — the grandson of the discoverer. Benford's contention and worry were about a base large enough to possibly disrupt the validity of Benford's Law. While it is accepted that the Benford phenomenon should be base invariant and independent of the choice of B, there is some limitation that one cannot ignore. If we consider some particular Benfordian data set in the context of base 10, then we expect the data set to also be base B Benford, assuming a base of a low value B similar to 10, such as 3, 6, 8, 13, 16, or 21. But if B is extremely large, then the data set will no longer be base B Benford and will lose its Benfordian status. A specific counterexample suggested by Benford is when the largest value in the data set is some integer M, and the smallest value is 1 (having no fractional values), and then the data set will not be base B Benford if, for example, $B = 2M$. Such a large choice of the base implies positive proportions of $\text{LOG}_{2M}(1 + 1/d)$ well beyond M and up to $2M - 1$, while the data set itself has zero proportions beyond its maximum value of M, and all this contradicts Benford's Law. However, if the data has significant portion of fractional data points below 1, then Benford's argument falls apart, since all sorts of first digits could emerge from that section of the x-axis from 0 to 1 of infinite digital cycles. Should Benford's Law be stated with the (troubling) qualification that the statistician must refrain from using very high base values? But if so, then how large is that limit? Would one-hundredth of the maximum be okay? Such a qualification would ruin the quality and universality of the law. If the law has to be stated differently for different data sets depending on the value of their maximum, then the law loses its base invariance status. This author suggests that the law implicitly focuses on practical and *natural B bases*, not exceeding 50, and definitely not over 100. Our positional number system and its associated base B is analogous to a measuring rod for physical quantities in nature or as a yardstick for real data. Imagine the manager of a small construction shop or the supervisor of some building material store constantly in need of measuring items that are priced depending on length, such as plastic board, plywood, metal pipes, long glass pieces, and cardboards. Prices typically read as, for example, "$37 per one meter of plywood" and "$1.48 per one centimeter of copper pipe". Without a good yardstick or a functioning measuring rod, no sale can take place. It is inconceivable that the manager would purchase a very long 25-meter yardstick that could hardly pass through the frame of the front door of the shop and which does not even fit nicely inside the establishment. The necessity to apply only a modest-size base B is mainly for our own number system itself and our own use of it in everyday life, for quantities, counts, and data, while the Benford necessity for it is merely a minor one.

Suppose that a very large base $B = 3000$ is chosen as the base for the positional number system for some odd civilization living on the edge of the Asian or American continent. Let us then imagine normal everyday quantities, such as the length of lands, houses, and tree logs, all in need of recording. Very few items are large enough for this base of 3000 to serve as a practical system. A rare exception would be some large land measuring 12,000, in which case the value is written as 40, namely $(4) \times (3000^1) + (0) \times (3000^0)$. Another lucky and rare quantity would be 9,006,000, which would be written as 120, namely $(1) \times (3000^2) + (2) \times (3000^1) + (0) \times (3000^0)$. All other regular everyday quantities below 3000 are written using a single symbol (digit) dedicated to that "low" value such as say 2759, hence in practice there exists no functioning number system for normal everyday quantities below 3000, while these "low" quantities are actually the ones most frequently in need of using in everyday real-life scenarios. Hence, in addition to the enormous difficulties of having to memorize 3000 "digits" or "symbols", such artificially high choice for the base precludes applying their number system for counting and calculating quantities below 3000, and this defeats the purpose of establishing a number system in the first place.

Empirical testing of real-life data sets and several abstract distributions with strong OOM values robustly confirm the base invariance principle in Benford's Law for a wide variety of bases ranging from 6 to 9000, and in spite of the above heuristic and specific arguments against choosing a base of very high value. For the US 2009 population census data of chapter 33, with 2.9 OOM of core data between the 3% percentile of 67 and the 97% percentile of 57316; and testing the set of seven Base value choices of {6, 10, 26, 50, 100, 160, 9000}, the corresponding set of seven SSD values are: {0.3, 1.3, 1.6, 9.6, 7.3, 12.3, 21.1} and which are fairly low, indicating conformity to Benford's Law. For the data set of weights in grams of all known 45,566 worldwide meteorite landings from outer space in all recorded history, downloaded from the website https://catalog.data.gov/dataset/meteorite-landings, and with 4.1 OOM of core data between the 3% percentile of 0.8 and the 97% percentile of 10000; the corresponding set of seven SSD values are: {1.1, 0.2, 1.0, 1.3, 1.4, 1.0, 9.1}. For 35,000 simulations from the Lognormal with location 3 and shape 2.7, and with 4.4 OOM of core data between the 3% percentile of 0.1 and the 97% percentile of 3170; the corresponding set of seven SSD values are: {0.1, 0.3, 0.4, 0.3, 0.2, 0.2, 3.1}. For random simulation of 100-kilogram rock breaking in 13 stages, namely one-dimensional random staged partition of chapter 18, and with 6.0 OOM of core data between the 3% percentile of 0.0000000865 and the 97% percentile of 0.086; the corresponding set of seven SSD values are: {0.5, 0.6, 0.8, 0.8, 0.9, 0.8, 2.6}. Hence, even for the extremely large Base value of 9000 we obtained very strong Benford behavior with the low SSD values of 21.1, 9.1, 3.1, and 2.6!

Yet, the base-invariance principle fails for data sets having lower OOM values which are just sufficient enough for minimal compliance with decimal Benford. A good example of this is the 2012 global earthquake data of chapter 5, with only 2.2 OOM of core data between the 3% percentile of 43.1 and the 97% percentile of 6216.6; where the corresponding set of seven SSD values are: {1.6, 3.1, 15.9, 31.1, 83.1, 42.7, 101.3}. For 30,000 simulations from the Exponential distribution with Lambda parameter of 7, and with only 2.1 OOM of core data between the 3% percentile of 0.0043 and the 97% percentile of 0.496; the corresponding set of seven SSD values are: {1.6, 8.9, 31.9,

69.5, 105.9, 128.0, 109.7}. These last two examples show decisive deviations from Benford for higher Base values, and which is due to their weak OOM data configuration.

Hence the generic conclusion here is that 'not all (decimal) Benford data sets are created equal', but rather those with a larger order of magnitude are by far more resistant to big Base values, sustaining their Benfordian status even under extremely large choices of the base such as 9000! While data sets with just sufficient OOM for decimal base 10 compliance with Benford (or for other low Base values) are fragile and they easily crumble as the choice of the Base value increases. All this is in perfect harmony and consistent with what was seen in the case of the CLT's Achilles' Heel of chapter 16, where data sets with a very large order of magnitude were shown to be by far more resistant to detrimental additive processes in the system.

Yet, considerable deviations from Benford are encountered even with moderate to higher OOM values whenever most of the data falls on the range of the x-axis between 1 and the Base. One such counterexample and exception is provided via Oklahoma State 2011 expenses list for the entire year encompassing 967,682 payments, and with 3.4 OOM of core data between the 3% percentile of $7.45 and the 97% percentile of $18,087.02; where the corresponding set of seven SSD values are: {1.9, 0.9, 3.5, 6.9, 11.5, 11.4, 57.2}. Here 77% of data points fall on the interval (1, 9000), and therefore the proportions of first digits are by far more equal and flatter compared with Benford proportions of Base 9000, as shall be further discussed in chapter 47. The author plans to do further research regarding the entire complex narrative encompassing (I) the choice of the base, (II) the spread of the core data on the x-axis including its minimum and maximum, and (III) the value of the OOM, in order to shed brighter light on resultant compliance with Benford's Law and to be able to state broader rules of expected compliance from data sets.

In any case, as one final confirmation of the decisive role played by the OOM regarding base invariance, two sharply contrasting Lognormal distributions shall be simulated with 35,000 realizations, one with extremely high variability and another with a very narrow focus. For simulations from the Lognormal with location 11 and shape 0.95, having the low 1.5 OOM of core data between the 3% and 97% percentiles; the corresponding set of seven SSD values are: {0.1, 2.4, 36.2, 83.9, 114.4, 265.0, 225.8}, hence, even though this simulated data deceitfully attempts to masquerade itself as Benfordian under decimal and other low Base values, its true non-Benfordian nature can be revealed by checking it against high Base values. For simulations from the Lognormal with location 21 and shape 7, having the very high 11.4 OOM of core data between the 3% and 97% percentiles; the corresponding set of seven SSD values are: {0.18, 0.31, 0.07, 0.18, 0.19, 0.16, 0.22}, hence this simulated data is decisively base invariant and truly Benfordian in nature. Note: the misguided measure of deviation known as MAD (Mean Absolute Deviations) applied to the above Lognormal(21, 7) and the seven bases, yields steadily decreasing values as the choice of base B increases: {0.00164, 0.00149, 0.00044, 0.00042, 0.00030, 0.00023, 0.00002}. This demonstrates the misguided conceptual basis of the MAD measure, as it wrongly divides the sum of absolute deviations by the number of digits (e.g. dividing by 9000 in the last base case). Here for base 9000, the misguided motivation of the MAD approach in attempting to counter the perceived but non-existent increase in total sum of absolute differences due supposedly to the large number of digits — is evidently not so. Here that fixed 100% total of theoretical is being divided and shared among 9000 digits, and each obtains only a tiny share of it, causing each difference to be of a tiny value. For example, Benford log base 9000 values of $(1 + 1/d)$ for digits {1, 2, 3, 80, 400, 2500, 3600, 5600, 8700} are {7.6%, 4.5%, 3.2%, 0.136%, 0.02742%, 0.00439%, 0.00305%, 0.00196%, 0.00126%} respectively, and which are of tiny values, therefore there is no need to divide total absolute deviations by 9000, as this will cause the measure to be excessively and artificially reduced.

Chapter 39

First Two Digits versus Last Two Digits

The probability of any first-two-digits (FTD) combinations, say 34 and exemplified in numbers such as 348, 0.03417, 3425.79, and 34.56, is given by the expression

$$P(p, q) \equiv \text{Probability[First digit is } p \text{ AND Second digit is } q]$$
$$= \text{LOG}(1 + 1/pq)$$

The term pq does not refer to the multiplication of digit p by digit q as in $(p) \times (q)$, but rather to the expression of the number $(p) \times (10) + (q) \times (1)$. For example, $P(1, 0) = \text{LOG}(1 + 1/10) = 0.0414$, which is of the highest probability; $P(2, 5) = \text{LOG}(1 + 1/25) = 0.0170$; $P(9, 9) = \text{LOG}(1 + 1/99) = 0.0044$, which is of the lowest probability. There are 90 FTD possibilities in total, namely $\{10, 11, 12, \ldots, 97, 98, 99\}$.

The Benford uniformity of mantissa on (0, 1), as in Figure 22, has its equivalency of 10^{MANTISSA} distribution of k/x type over (1, 10), as in Figure 25. FTD combinations, such as 25, lie exclusively in the subrange of (2.5, 2.6) within the (1, 10) range. Converting this subrange into a mantissa subrange via the solutions to $2.5 = 10^{\text{MANTISSA2.5}}$ and $2.6 = 10^{\text{MANTISSA2.6}}$ points to the span on the mantissa-axis in Figure 22 between LOG(2.5) and LOG(2.6), hence the probability is $\text{LOG}(2.6) - \text{LOG}(2.5) = \text{LOG}(2.6/2.5) = \text{LOG}(26/25) = \text{LOG}((25+1)/25) = \text{LOG}(1 + 1/25)$. More generally, for any pq combination, the algebraic expression is

$$P(p, q) = \text{LOG}(pq + 1) - \text{LOG}(pq) = \text{LOG}((pq + 1)/pq)$$
$$= \text{LOG}(pq/pq + 1/pq) = \text{LOG}(1 + 1/pq).$$

The probability of any first-three-digits combinations, say 374 and exemplified in numbers such as 37428, 0.0374315, 3741.29, and 37.4653, is given by the expression

Prob[First digit is p AND Second digit is q AND Third digit is r] = LOG$(1 + 1/pqr)$.

The probability of any last-two-digits (LTD) combinations, say 59 and exemplified in numbers such as 21759, 0.041659, 18159, and 2000.59, is simply 1%, and this is so equally for all possible LTD combinations. The reason for this result is that, typically, numbers in real-life data sets are not too short digit-wise but are rather sufficiently long, i.e., there are normally plenty of significant digits within each number, say over 4 or over 5 digits, hence the LTD combinations typically pertain to the high third, fourth, or fifth digital orders, or even higher orders, and consequently each digit — on or near the right side of numbers — is of an approximate equal probability of 1/10, or 10%. Since beyond the first and second orders, higher-order digits are nearly independent of each other, with 10% each regardless of other orders, it follows that the probability of the combination of two digits is simply the product of each probability, namely the equal probability of $(1/10) \times (1/10)$ or 1/100 for each possibility, so 1% per pair of digit combination from the 100 total possibilities of $\{00, 01, 02, \ldots, 97, 98, 99\}$. For example, the probability of the LTD combination of 59 is 1%, assuming all the numbers in the Benford data set are with lots of digits per number. The charts in Figures 34 and 35 depict the histograms of FTD and LTD combinations according to Benford's Law. These two charts are essential tools in digital forensics in the context of data fraud detection.

When a number is blessed with lots of significant digits, such as 96,378.25, it obviously contributes 96 for the FTD test and 25 for the LTD test.

For shorter numbers, such as 39, 239, or 839, the LTD combination is indeed 39, since the "last" digits are counted from the right-most side, yet 39 is not of the equal 1% probability, as the digit 3 here pertains to the first digit or to the second digit, which are of unequal probabilities, and not of the third, fourth, or higher orders with equal probabilities.

Non-zero single-digit numbers, such as 8 and 0.006, should be incorporated only in the first-order test and eliminated altogether before performing the second- and third-order, FTD, and LTD tests. That number 8, which does not possess a second digit or a third digit, should not be artificially and forcefully converted into 8.00, and pretending that its second and third digits are 0 is a fallacy, since they simply do not exist here. A two-digit number, such as 59, should be incorporated only in first, second, and FTD tests and eliminated before performing third and LTD tests. The number 59 should not be artificially and forcefully converted into 59.00.

Alas, usually, the forensic digital analyst can easily decide on things based on the decimal precision point of the data under consideration. For dollar amounts with two decimal points of precision for the cents, a five-dollar sale or an expense figure written as \$5 is indeed 5.00, and its second and third digits exist and are indeed 0. For dollars, all amounts of \$0.09 and below are excluded from the second and higher orders, FTD, and LTD. All amounts of \$0.99 and below are excluded from the third and higher orders as well as LTD. Only amounts over \$9.99 may contribute to LTD, especially amounts over \$99.99. For count data with 0 decimal points of precision, such as population data, a city with 9 inhabitants or less contributes only to the first order, as it does not possess

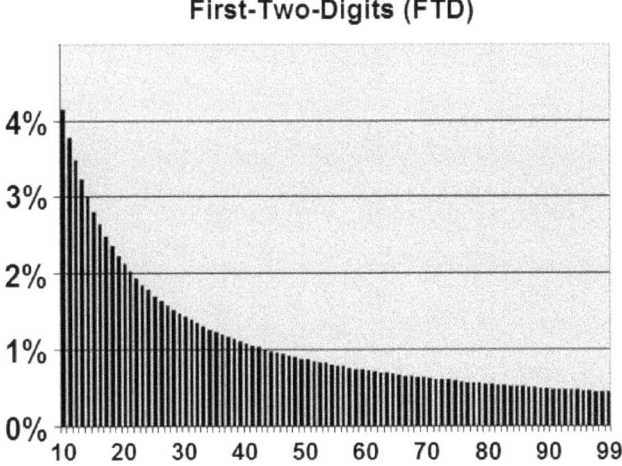

First-Two-Digits (FTD)

Figure 34. Benford's Law: Chart of First-Two-Digits Combinations

Last-Two-Digits (LTD)

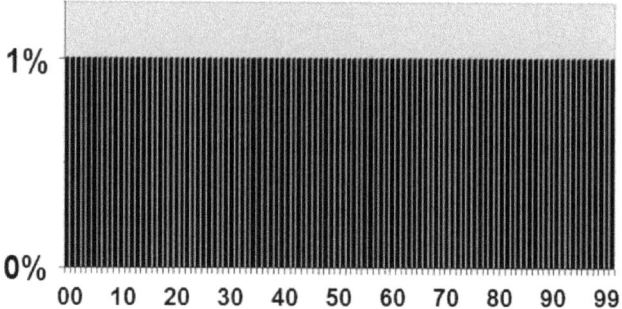

Figure 35.   Benford's Law: Chart of Last-Two-Digits Combinations

second or third digits and should never be artificially written as "9.00 inhabitants". A city with 10–99 inhabitants contributes only to the first, second, and FTD tests, not to the third order or LTD. Only a city with more than 999, and especially over 9999, inhabitants may contribute to the third order and LTD.

SECTION V
The Law of Relative Quantities

Chapter 40

The Related Concepts of Digits, Numbers, and Quantities

Benford's Law, a supposedly digital phenomenon, is found and demonstrated to be the consequence of a more general and universal quantitative law, coined "the law of relative quantities". The focus then naturally shifts from the strong attention given previously to the relative occurrences of digits in numbers within data, to the investigation of the relative occurrences of quantities via the shape and structure of histograms of data. Physical quantities in the real world are represented by writing numbers for them, and these numbers are written by way of our digital language; therefore, in that sense, digits ultimately represent physical quantities. The Benford phenomenon originates from the fact that, in extreme generality, nature creates many small quantities, but only a very few big quantities. It follows that Benford's Law is applicable just as well to data written in the ancient Roman, Egyptian, and other digit-less civilizations, without insisting on the conversion of their values into our decimal positional number system so that the physical phenomenon could manifest itself; rather, the phenomenon could manifest itself directly in the structure of the histograms of Roman or Egyptian data. This more general law regarding quantities should be stated independently of any arbitrary societal number system and digits.

Chapter 41

The Arbitrariness of Our Positional Number System

Our positional number system is extremely efficient, but it's still arbitrary. It might possibly be the most efficient number system in existence, and there may not be any superior or more efficient one, yet we should acknowledge its arbitrariness. We are so used to reading, writing, calculating, and working with numbers in this system from a very young age that we tend to associate them with something "essential" or "absolute", and we might believe that our numbers are the only natural and proper way to express quantities. Other number systems seem to us as being "game-like" and serving merely as an "intellectual exercise". We also tend to think of our number system as imperative, and some might imagine that the entire mathematical edifice built over the past two and a half millennia is based on it. But this is not true. The creative Greeks of antiquity did not have our positional number system, yet they excelled in mathematics and geometry. We need to break out of this subtle mathematical orthodoxy and dogma. Indeed, we could have constructed all our mathematics using, say, Roman numerals, and our sciences could also have been built and based on any primitive and inefficient number system, and the only price we would have paid is inefficiency and slow calculations necessitating more mental effort, but the mathematics and the sciences would be there just fine.

Hence, Benford's Law, being so intimately involved with our number system to the extent that it is stated in terms of its symbols (digits), is arbitrary just as well. This realization leads one to suspect that $LOG_{10}(1+1/d)$ for the first digits does not account for the main essence of the phenomenon and that there exists possibly a more universal and non-arbitrary law.

It is necessary to remind ourselves of the distinction between the verbs "discover" and "invent".

We have *discovered* that the sides of a right triangle relate, as in $C^2 = A^2 + B^2$.

We have *discovered* Newtonian physics and that $F = M \times A$ and $F_G = G \times M_1 \times M_2 / R^2$.

We have *invented* a very efficient number system, the positional number system base 10.

Chapter 42

Two Radically Different Interpretations of the Benford Phenomenon

Two radically different interpretations of the Benford phenomenon are given:

First: This is truly a physical phenomenon existing independently of us, our number system, and our way of recording data; it is a quantitative and a physical law of nature.

Second: This digital pattern found in physical data is simply due to our own peculiar system of counting values by way of their digital representations; hence, the phenomenon is all about our number system itself and not about the natural world. Consequently, the phenomenon is arbitrary and has no independent existence outside our digital perception.

As an analogy for the second interpretation, a child wearing red eyeglasses may believe that every physical object in the world is red and would ask his or her father, "Daddy, how come everything in the world is red?" The red color on the eyeglasses is arbitrary, and the fact that everything appears red is arbitrary as well. Had the child been wearing green eyeglasses, everything would have appeared green. According to this interpretation, the Benford digital pattern found in physical data is simply due to our own peculiar way of viewing values via our unique digital prism. Figure 36 depicts the mathematician or scientist observing a quantitative and

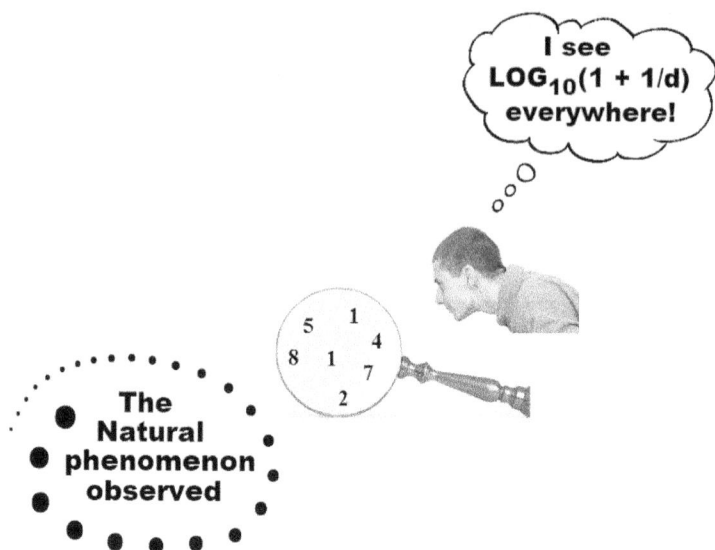

Figure 36. The Scientist is Observing Benford's Law via our Digital Lens

physical phenomenon and noticing that its digital structure within our positional number system is in compliance with Benford's Law — as in the second interpretation.

Should we expect to find any first numeral pattern in data found in the ancient archives of the Roman Empire written in their Roman numeral number system? Could we perhaps find any pattern in our modern-day data when converted to Roman numerals? The answer is decisively in the negative, and this is so because of profound theoretical reasons relating to the odd, uneven, and irregular way the cycles or occurrences of these numerals are spread over the x-axis, and not because of the inefficiency of this ancient system per se. In addition, empirical testing of Roman numeral data regarding any first numeral pattern fails to show any consistent pattern across a variety of data sets, as each individual Roman data set yields its own distinct proportions of first numerals. Figures 37 and 38 depict one part of the failed attempt to find any first numeral pattern in the Roman numeral number system.

DLXXXV	CCLXXX	XVII	XCII
CDIV	MVIII	CDXLIII	VCIII
DCLXXVII	LDXV	LXXVII	CMVLIII
LMXLIII	DCCX	CCCVCIV	LXVII
DLXVI	LXXII	XLIV	LII
CCIX	MCMXI	CDV	CXXII
DCIX	LDXII	XXVIII	CDXLIII
DCVC	LXVII	CIX	DCLV
MMCLIII	IV	XXVI	CCCLXVI
DCII	LXXXIV	CCCXXI	XXXII

Figure 37.   An Ancient Data Set from the Archives of the Roman Empire Era

DLXXXV	CCLXXX	XVII	XCII
CDIV	MVIII	CDXLIII	VCIII
DCLXXVII	LDXV	LXXVII	CMVLIII
LMXLIII	DCCX	CCCVCIV	LXVII
DLXVI	LXXII	XLIV	LII
CCIX	MCMXI	CDV	CXXII
DCIX	LDXII	XXVIII	CDXLIII
DCVC	LXVII	CIX	DCLV
MMCLIII	IV	XXVI	CCCLXVI
DCII	LXXXIV	CCCXXI	XXXII

Figure 38.   The First Numerals of Data from the Roman Empire Era

Chapter 43

The Quest for a Universal and Number-System-Invariant Measure

Imagine two observers on two distinct planets A and B, both clearly observing a distinct quantitative phenomenon occurring on a third planet C via powerful telescopes, as depicted in Figure 39. Planet B shown on the right-hand side has invented a positional number system base 10 and thus observes Benford's Law very clearly, where the first digits are nearly as in $\text{LOG}_{10}(1 + 1/d)$ for the data set of planet C, as well as for almost all other data sets on its own planet. Planet A, shown on the left-hand side, has invented Roman numerals and is numero-conservative, emotionally attached to its number system and adamantly refusing to adopt any other numerical system. Planet A proclaims that it does not see anything particular in the data set of planet C, and it is asking for help in constructing a quantitative and number-system-neutral measure that would yield a consistent pattern for the planet C data as well as for almost all other observable data sets on its own planet.

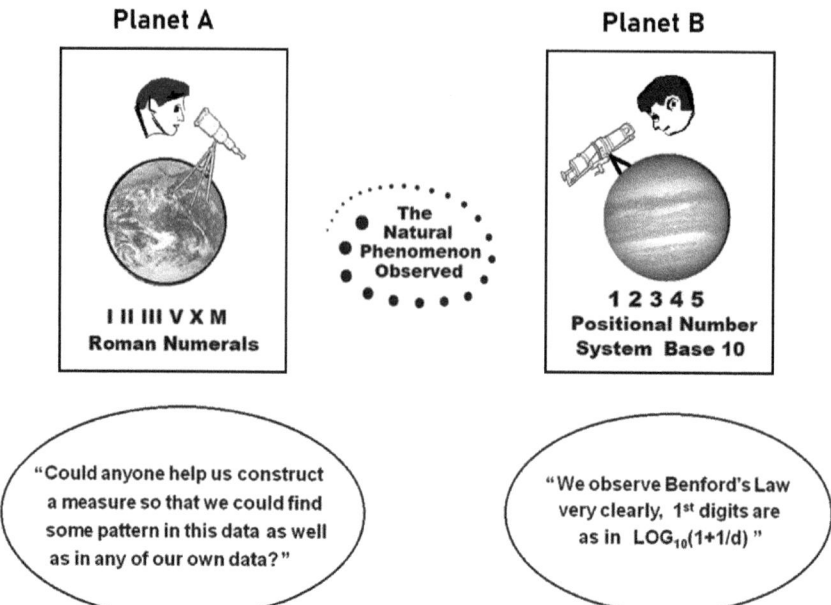

Figure 39.   The Benford Dispute between Planets of Distinct Number Systems

Philosophically, and according to the first (and correct) interpretation of the Benford phenomenon, it is necessary that both planets come up with a universal, primitive, and invariant mathematical measure agreed upon by all for this clearly and easily observable physical phenomenon on planet C, since the phenomenon exists in its own right independently of any observers out there and independently of any arbitrarily invented number-system. In other words, a singular quantitative statement should be formulated, which would be identical for all observers, being number system invariant.

But what measure could it be? Naturally, the idea that comes to mind here is that that universal and primitive measure to be agreed on by all observers could be a mathematical expression relating to the commonly observed histograms of the data sets on planet C, since the shapes of histograms occurring on planet C are seen identically on planets A and B.

But what aspect of the histogram could be measured and proven to be the universal pattern? One characteristic common to almost all data and which leads to the Benford phenomenon is the positively skewed histogram, falling on the right for high values. This implies having many small values but only a very few big ones. This is a universal feature, being number system invariant. Therefore, a precise quantitative measure of such a fall in histograms may perhaps serve as a general law, true for all observers regardless of their number system in use.

Hence, let's change the agenda, and instead of focusing on digit proportions within data sets, let us focus on histograms of data sets and their structure.

Chapter 44

The Shape and Nature of Histograms are Number-System Invariant

Facilitating this generalization is the fact that histograms do not depend on the number system in use. The histogram of any data set viewed through the prism of several distinct number systems appears totally identical. Histograms apply the Cartesian X–Y plane, which is, in essence, a purely geometric construct, where units of length on the horizontal X-axis signify the units of the physical quantity being measured (in essence, counts, such as how many kilos or how many meters), while the units of length on the vertical Y-axis signify the counts of quantities falling within certain bin limits, and these counts are neutral regarding any numerical language. Therefore, the visual aspect of a histogram, the relative heights of the bins, and so forth, are all fixed and invariant with respect to the number system in use — well, except for the numerals, numbers, digits, or whatever symbols that are written along the horizontal and vertical axes, which are indeed dependent on the number system. Clearly, the message conveyed in a given histogram is universal, irrespective of the number system in use, hence the "histogram number-system invariance principle". Since statistical density distributions are simply the continuous forms of discrete histograms (infinitely refined), the principle is very general, and all statistical density distributions and curves are number-system invariant as well.

In order to highlight and visualize the fact that histograms do not depend on the number system in use, and that the visual form or the shape of a histogram does not change when we switch to another number system, four histograms written in four distinct numerical languages (but all standing for and representing the same identical data set) are depicted in Figures 40–43.

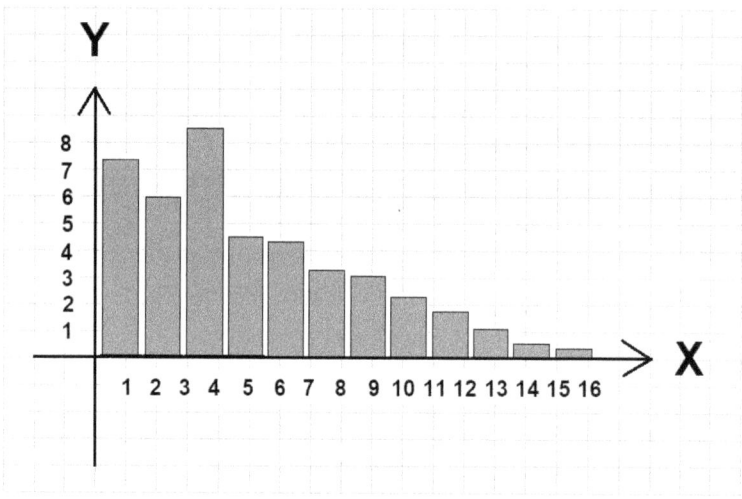

Figure 40. Histogram Appearance with Positional Number System Base 10

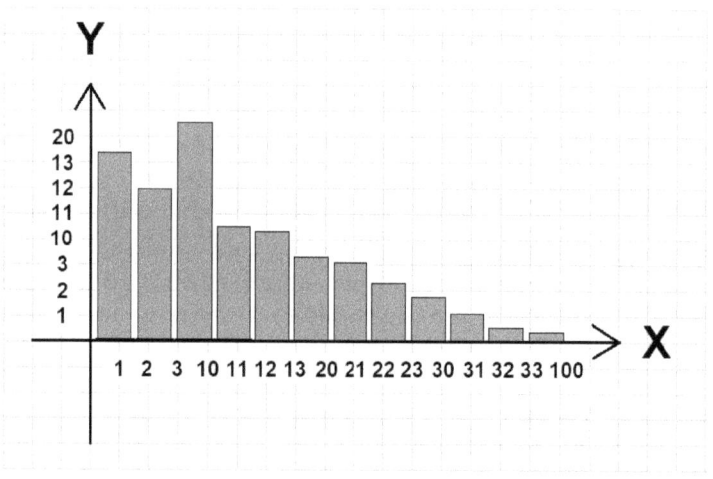

Figure 41. Histogram Appearance with Positional Number System Base 4

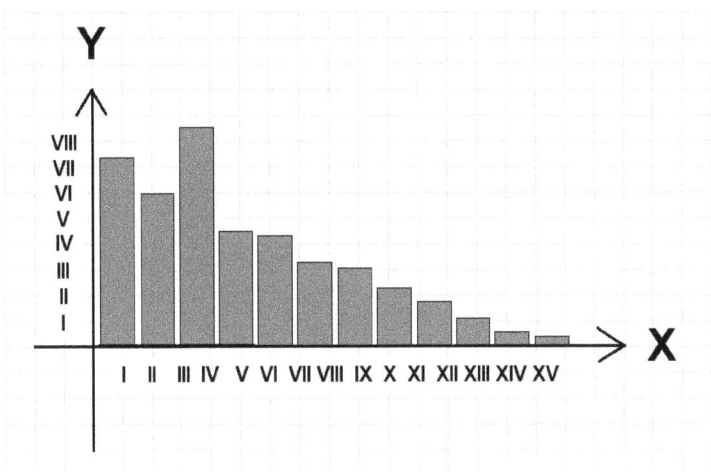

Figure 42. Histogram Appearance with Roman Numerals System

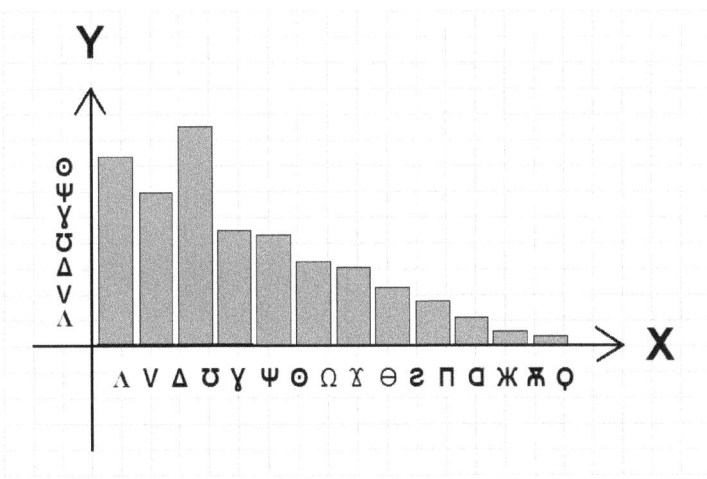

Figure 43. Histogram of a Primitive Society Absent a Number System

Chapter 45

Constructing a Three-Bin Histogram Signifying Small, Medium, and Big

In attempting to discover a universal quantitative measure, we construct a singular histogram with only three bins of equal width, signifying "small", "medium", and "big", tailor-made to individually fit each data set by spanning the entire range of its data from its minimum to its maximum, all in order to detect overall skewness, as depicted in Figure 44. We record the relative proportions of the three bins for each data set and observe that, indeed, nearly all data come

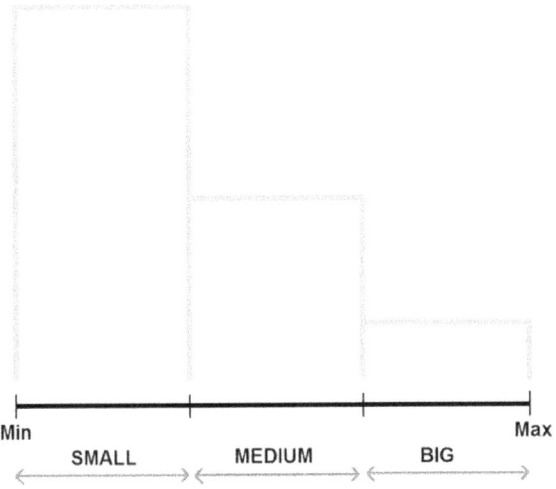

Figure 44. Crude Three-Bin Histogram Scheme Measuring Relative Quantities

with larger "small" proportions than "big" proportions, i.e., the first bin relating to "small" is much higher than the third bin relating to "big".

Unfortunately, when empirically checked against a large collection of real-life data sets, this scheme leads to a decisive failure. Yes, indeed, positive skewness is detected for nearly all data sets, albeit without an exact or consistent measure. The first-ranked bin for "small" varies widely, fluctuating between 50% and 95% approximately, while the third-ranked bin for "big" varies widely as well, fluctuating between 1% and 15%. This demonstrates that each data set possesses a dissimilar numerical measure of skewness when (wrongly) measured in such a crude manner. This decisive failure to obtain a consistent numerical measure, where a steady pattern is not found across all data sets via a singular three-bin histogram, calls for a radical change in our approach regarding how relative quantities and sizes should be measured by way of histograms, so that a nearly universal numerical result could be found.

There are two intrinsic features in the above three-bin scheme that need to be abolished. The first feature is having a shifting histogram, constructed to fit the particular data set on hand by aiming at the particular range of the data, from its minimum to its maximum. Each three-bin histogram in the above scheme is made to span different parts of the horizontal x-axis, depending on the data set under consideration. The second feature is lazily constructing only one global histogram, attempting to calculate relative quantities and sizes in one fell swoop, instead of laboring long and hard by constructing a large set of detailed and repeated local mini histograms, all with the same fixed and identical number of bins within each histogram, which could then be aggregated for each bin rank as one singular result. The histograms should also be constantly expanding and widening in length and scope. The motivation for such a construction utilizing numerous mini histograms is to keep checking local relative quantities in all parts of the x-axis and, by implication, checking within all subranges of the data, taking their distinct quantitative pulses everywhere along the x-axis, followed by the simple aggregation of all these pulses into one decisive result.

Hence, instead of constructing a singular histogram with three bins, constructed subjectively and differently for each particular data set, what is necessary here is an objective, autonomous, fixed, and universal set of numerous histograms to be positioned onto the x-axis from zero onward to positive infinity. Such a common set of numerous histograms should be constructed for use in all data sets, unifying our approach to them in a sense, with the hope of achieving a measure of universality and consistency. A clue to this approach in measuring the fall in the histogram was obtained by simply imitating and generalizing the histogram structure hidden within the digital arrangement of Benford's Law, since the law subliminally or subtly refers to infinite numbers of histograms, starting at the origin 0, each with nine bins (for the nine first digits) and repeatedly expanding in size by a factor of 10 (the base 10). In essence, Benford's Law gave us the clue for the bin structure of the general law of relative quantities. It was reverse engineering of sorts.

Benford's Law itself was not discovered by someone who sat down deep in thought, inquiring about the supposed distribution of digits within numbers in real-life data and attempting to arrive at an exact mathematical expression in the abstract. Most likely, nobody in all of human history ever contemplated the issue, and almost certainly nobody ever actually empirically examined digit distributions within data before all these were discovered by Newcomb and Benford. Rather, the law "forced itself upon us" by gently flaunting its physical manifestation, giving us the clue to the very existence of the phenomenon via the differentiated physical wear and tear of pages in old books of logarithmic tables.

For a deeper understanding of Benford's Law, it is necessary to relate and connect the occurrences of d as the first digit of the number X to the position of X within the x-axis. Knowing the precise location of X on the x-axis specifically determines the first digit of X, but knowing the first digit of X does not determine its location on the x-axis. For example, knowing that the first digit of X is 8, number X is still very much unknown: it could be 8703, or 0.0085, or 81, or 846, and so forth. Figure 45 demonstrates the locations on the positive x-axis where digit 1 leads; but this is missing the infinite

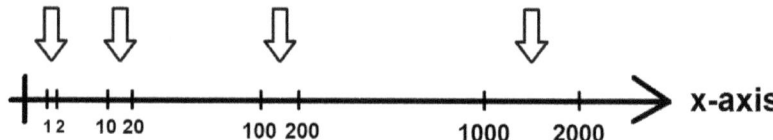

Figure 45.   The Locations of the Subintervals on the *x*-axis where Digit 1 Leads

subintervals below 1.0 of small fractional values, such as 0.0017 or 0.13; and this is also missing the infinite subintervals beyond 10,000 of big values, such as 10,457 or 18,490,327. This also demonstrates the cyclical way first digits occur on the *x*-axis and the fact that these cycles keep expanding, forever widening their span.

Figure 46 relates to the real-life data on the US population regarding all its 19,509 cities and towns in the 2009 census. The figure demonstrates clearly and visually the vista of Benford's Law as a set of expanding histograms, forever widening their widths. The entire or effective first-digit distribution of a given data set is nothing but the condensed or aggregated histogram of all the various nine-bin local mini histograms standing between the integral powers of 10, such as 1, 10, 100, and 1000. But this is done bin by bin, digit by digit, separately aggregating each digit's uniquely ranked bins within all of these infinite sets of histograms. Incidentally, Figure 46 also visually and decisively demonstrates the digital development pattern for the US population data, where the high digits are actually more frequent on the left for small values, albeit for a very small portion of data points, so this hardly affects the overall Benfordian configuration of the entire data set. Yet, the triumph of the high digits there is fleeting, as this is followed by a complete reversal and the dominance of low digits further to the center and even more so to the far right for big values; however, this topic is not related to the general quantitative law described in this section.

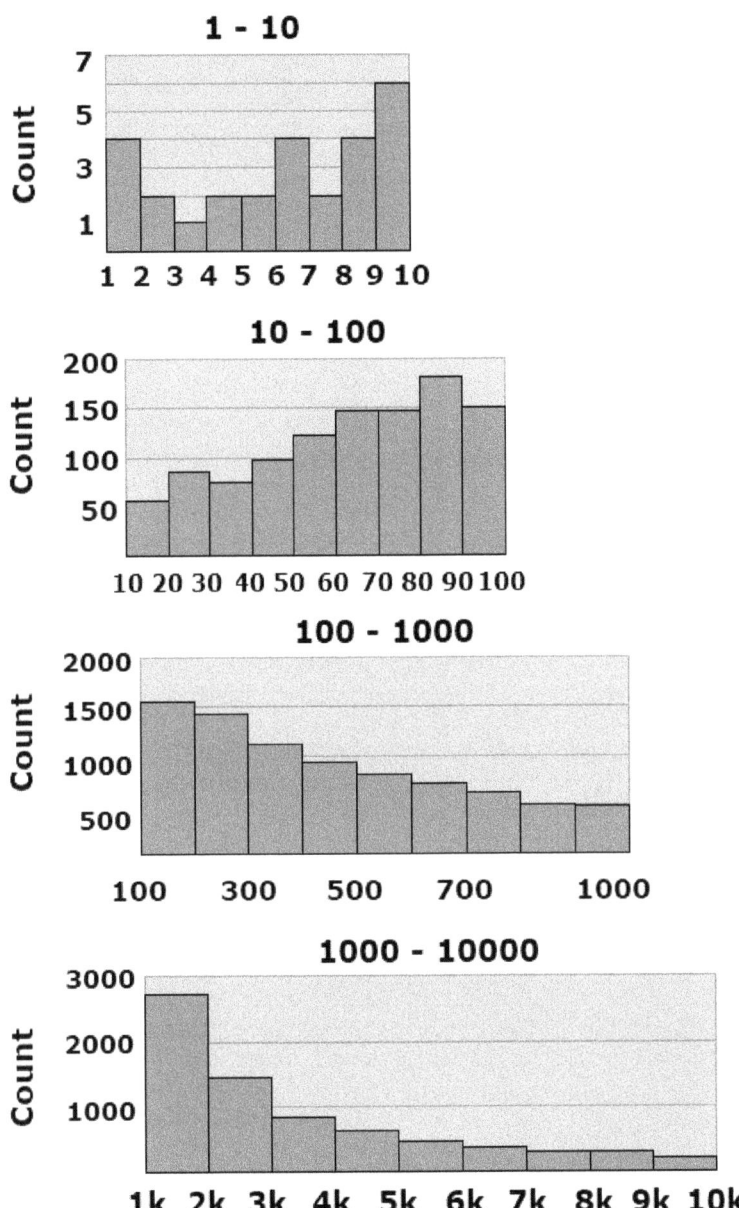

Figure 46. The Benford Vista of Expanding Histograms in Pop. Data

Chapter 46

Constructing a Set of Infinitely Expanding Histograms

We construct a set of repeated local histograms from the zero origin to positive infinity, followed by the aggregation of all these histograms into one singular result.

Each local histogram is wider than its adjacent one to its left, i.e., the widths of the histograms and their constituent bins are always expanding, inflating in size, doubling, tripling, or increasing by some multiplicative factor F greater than 1.

The need to keep expanding the width of the histograms is due to some deep-rooted statistical reasons related to the heuristic argument that typically values in real-life data spring from numbers that are "rapidly expanded" and "stretched out" along the x-axis, getting diluted on the right for big values, hence a slow system with identical histograms having fixed bin widths would not be able to "keep up" with such a rapid pace of the outlay of the data itself and could not serve as an appropriate measuring arrangement. Consequently, our measuring net of mini-histograms should indeed expand just the same. If the *measured* data accelerate quantitatively, then the *measuring* bin scheme should also accelerate accordingly.

Figure 47 depicts the sketch of one such construction of repeated three-bin histograms expanding by a factor of 2 on each cycle, showing only the x-axis ranges of the bins regarding the first three histograms (cycles) in the whole scheme and neglecting to show the rest of the (infinitely) numerous and much wider histograms on the right for brevity and lack of space.

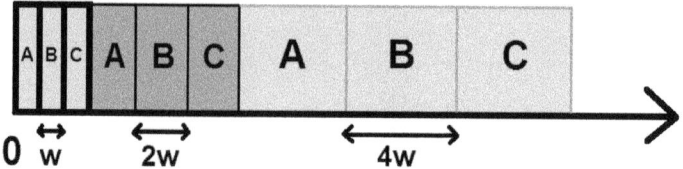

Figure 47.   Three-Bin Histograms with 2 as the Factor of Expansion

In the scheme shown in Figure 47, each histogram comes with an equal and fixed number of three bins, called A, B, and C, all having the same bin width. The sizes of the histograms are expanding (doubling) with each cycle, so that their bin widths are different for each histogram; some are very short and narrow (on the left), while others are very long and wide (on the right). The overall goal is to measure the relative proportions of each bin, A, B, and C, by comparing the total data points falling within all bins ranked A, the total data points falling within all bins ranked B, and the total data points falling within all bins ranked C. The separate aggregation of the three bins of the histograms, bin by bin, tells us indirectly about the overall shape of the histogram as well as about relative quantities for the data set under consideration in its entirety. If the data is positively skewed and the histogram falls to the right overall, then the total data points falling within all bins A (which always pertain to relatively smaller quantities) should be greater than the total data points falling within all bins B (which always pertain to slightly bigger quantities). And the total data points falling within all bins C (which always pertain to even bigger quantities) should be less than the total data points falling within all bins B.

The term "**relative quantities**" reflects the fact that sizes are relative. Planet Earth is very small relative to the Sun but big relative to the planet Mercury. Planet Earth is huge relative to the small city-state of Singapore. Had the entire Universe consisted only of Earth and nothing else and without any inner divisions or marked parts, then one could not determine whether Earth is big or small, since the terms "big" or "small" are not absolute but relative. For data falling within the range (200, 700), the part of (600, 700) is relatively big (and the rarest), but for data falling within the range (600, 3000), the part of (600, 700) is relatively small (and the most frequent).

More generally, the scheme is of repeated cycles of local histograms with D **number of bins** each, where the histogram width is expanding at **an inflation factor of** F, followed by the aggregation of all the bins of the histograms into a singular set of proportions. In Figure 47, the arrangement is that of two parameters, $D = 3$ and $F = 2$. The entire system starts from origin 0 with an initial infinitesimally small size of bin width w. In the second cycle, the width of each bin is $2w$. In the third cycle, the width of each bin is $4w$. In the fourth cycle, the width of each bin is $8w$, and so forth. The notation used is the lowercase w, standing for and hinting at the first letter in the word "width". We also denote lowercase d as the bin rank with respect to D, with d running from 1 to D so that $d \in \{1, 2, 3, \ldots, D\}$. In the particular example in Figure 47, d runs from 1 to 2 and then to 3, namely, from the first bin A to the second bin B and then to the third bin C.

How small should the first width w be made in practice so that statisticians and data analysts could actually implement a computer program to measure real-life data sets without worrying about highly abstract and complex mathematical concepts, such as "infinitesimal" and "limits"? A straightforward and practical rule of thumb is to choose the value of w as, say, one-millionth or 0.000001, i.e., 10^{-6}, assuming that this small value is less than one-hundredth of the minimum of all the minimums in the collection of the data sets to be examined. It must be noted that as long as w is made to be exceedingly small, its exact value does not affect the results to any significant degree whatsoever. In other words, empirical testing with real-life data shows that results are practically the same for whatever value is assigned to w, so long as w is made to be exceedingly small. To recap, results are practically independent of w and only depend on D and F. Also of note is that even though the entire system should start by default from the 0 origin, moving the starting point a bit forward does not affect results to any significant degree whatsoever. Empirical testing with real-life data shows that results are practically the same for whatever starting point, so long as it is exceedingly small and thus quite near the 0 origin. In addition, the scheme should be made to automatically stop whenever the maximum of all the maximums in the collection of the data sets to be examined is encountered by the computer, since there would be no more values to incorporate beyond that point, and the elusive goal of reaching "infinity" should be relaxed.

For each individual histogram, we carefully measure relative quantities within that histogram, namely, the number of data points falling within bin A, within bin B, and within bin C (or more), and then at the end, we aggregate this measure coming from all the histograms into one final result. In other words, at the end, we calculate the total of all the data points falling within all bins A of all the histograms by adding them, which is called $\sum$ bin A. Then, this is followed by the same calculations for all bins B, then for all bins C, and so forth, for all the bins, resulting in the aggregated vector ($\sum$ bin A, $\sum$ bin B, $\sum$ bin C, and so forth). Using the notation d for bin rank with respect to the number of bins D, this aggregated vector can be expressed as ($\sum$ bin rank $d = 1$, $\sum$ bin rank $d = 2$, $\sum$ bin rank $d = 3, \ldots, \sum$ bin rank $d = D$).

This entire arrangement shall be termed the "**bin scheme**".

Chapter 47

Numerical Consistency in Bin Schemes for 15 Real-Life Data Sets

Let us consider the application of the bin scheme to 15 data sets and abstract mathematical constructs. Detailed descriptions of these data sets are outlined in Appendix B. The bin scheme is made with five bins for each histogram, starting from the origin 0, constantly expanding by a factor of 2, having an initial very small bin width w of 0.00000000003, and automatically stopping whenever the bin width reaches 40,000,000,000,000. Figure 48 visually depicts the numerical results of the bin scheme. Each data set or case is depicted as a singular histogram expressing the result of the bin scheme. For each histogram of a given data set, with each successive bin, we alternate color from black to gray, to black, to gray, and then to black again. Staring at Figure 48, we conclude that the bin scheme has indeed succeeded superbly here! The average for these 15 data sets is {26.2%, 22.1%, 19.2%, 16.9%, 15.5%}, and the individual deviation for each data set from this overall average is very mild. A remarkable consistency and nearly an exact pattern have been found here across a variety of disparate data sets and cases. Several other bin schemes with different D and F values performed on these 15 data sets and cases also yielded a consistent pattern in how bin proportions fall to the right, each with a positive skewness, manifesting the "small is beautiful" principle.

The goal of the scientist or the mathematician is to avoid the need to state numerous laws for all types of special cases: a law for the particular case of $D = 5$ and $F = 2$ for example, and another law for

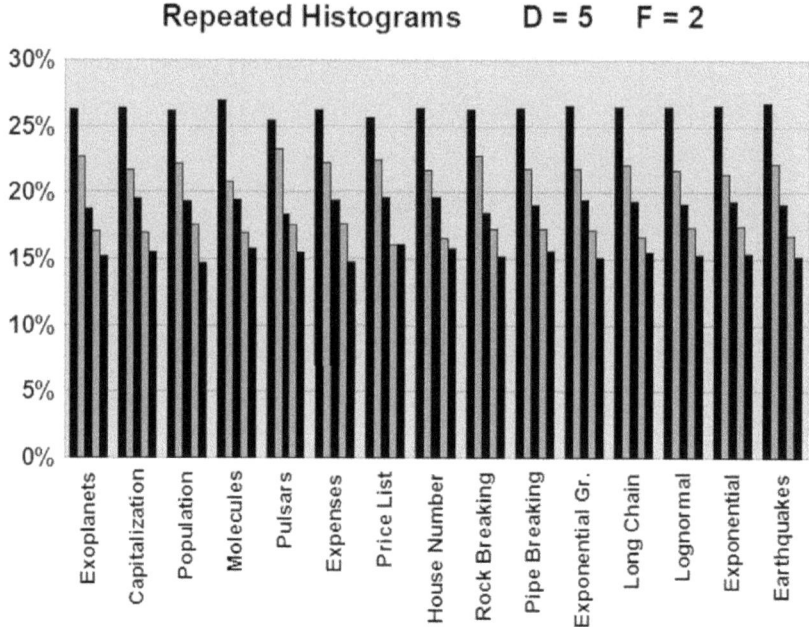

Figure 48. Pattern Found in 15 Data Sets: Bin Scheme $D = 5$, $F = 2$

the particular case of $D = 8$ and $F = 3$, and so forth. The challenging quest here is to find the generic mathematical expression that would predict the proportions of bin schemes for any given combination of D and F values in one fell swoop. Further encouragement supporting and motivating this quest is found in the fact that this nearly exact pattern of bin schemes is also confirmed in numerous other real-life physical and scientific data sets, far beyond the confines of the above 15 data sets and cases.

One possible approach in attempting to arrive at a generic mathematical expression is by empirically testing numerous data sets with a large collection of bin schemes for a wide variety of D and F combinations, creating a long list of D and F values along with their associated bin proportion results, and then attempting to find out **inductively** what is the best or most fitting mathematical expression as a function of D and F.

As one concrete example of the inductive approach, a variety of bin schemes are explored for the earthquake data set using several combinations of D and F values, as shown in Figure 49.

This data set pertains to the time in units of seconds between all successive earthquakes worldwide for the year 2012, totaling 19,452 earthquakes. Since the minimum value in the earthquake data is 0.01 seconds, the value of 0.002 was assigned for the initial bin width w in all of these bin schemes so as to make sure that all the schemes have a small and refined initial bin width, well below this minimum value of 0.01. Also, all bin schemes start from the origin 0.

The results in the table depicted in Figure 49 show that positive skewness is found in all of these bin schemes, as proportions are monotonically decreasing for higher-ranked bins for all D and F combinations, and this fits harmoniously with the fact that the small outnumbers the big in this earthquake data set, as can be empirically verified by examining its histogram.

Unfortunately, no straightforward hint about some obvious mathematical expression can be deciphered from the table in Figure 49, at least not for this mortal, the author, with limited induction experience. Philosophically, discovering the law

D	F	Bin A	Bin B	Bin C	Bin D	Bin E	Bin F	Bin G
3	2	41.3%	32.1%	26.7%				
3	3	46.1%	31.0%	22.9%				
3	4	50.0%	29.9%	20.1%				
3	5	53.1%	27.8%	19.1%				
3	6	55.9%	26.4%	17.7%				
3	11	63.6%	22.1%	14.3%				
4	2	32.1%	26.7%	22.2%	19.0%			
4	3	37.2%	26.5%	19.9%	16.5%			
4	4	41.4%	25.9%	18.4%	14.3%			
4	5	43.3%	24.6%	18.0%	14.2%			
4	12	53.5%	21.0%	14.9%	10.7%			
5	2	26.7%	22.2%	19.0%	16.9%	15.2%		
5	5	36.5%	22.5%	17.0%	13.3%	10.6%		
5	9	42.0%	22.5%	15.6%	11.1%	8.8%		
7	2	19.0%	16.8%	15.2%	13.6%	13.1%	11.5%	10.7%
7	4	25.9%	18.4%	14.3%	13.0%	10.6%	9.8%	8.0%
7	8	32.6%	19.0%	13.5%	11.1%	9.3%	8.0%	6.4%

Figure 49. Bin Schemes with a Variety of D and F Values: Earthquake Data

inductively is not the best approach since the inductive method is very difficult to implement and because it is prone to errors and uncertainty. A superior approach is to argue this by way of a conceptual postulate, which would then lead to the mathematical expression **deductively** — all the while closely agreeing with empirical results from real-life physical data sets.

Figure 49 also decisively demonstrates the fact that higher F values correlate with more skewness for any given fixed D value. Higher inflation factors F are associated with faster expansions of the histograms, thus resulting in a greater disparity between data falling in the first bin and data falling in the last bin. In other words, higher F values with stronger and faster expansions are correlated with greater skewness between the bins. This empirical result fits very well with another empirical fact, namely that any flat $F = 1$ bin scheme without any expansion whatsoever in the histograms and constructed for Benfordian data sets with skewness and high OOM yields bin equality in the approximate, that is to say, the same bin proportion of $1/D$ for all the bins, in spite of the skewness of the data itself. Evidently, as conjectured earlier via the heuristic argument in the previous chapter, without expansion of the histograms, the bin scheme is not capable of measuring data properly.

In addition, this result also fits very well with another empirical fact, namely that data (with or without strong OOM) spanning the x-axis range from 1 to Max is not Benford for very high Base value choices such as $B = $ Max, and relative bin results are more equal and flatter as compared to the theoretical Benford configuration which comes with more skewness as discussed at the end of chapter 38. Here F is not even relevant since there is only one long histogram with just one set of D bins and without any additional histogram cycles, precluding any expansions further to the right beyond the first single histogram, and leading to flatter configuration with less skewness.

Chapter 48

The Postulate on Relative Quantities

The Postulate: The generic pattern in how relative quantities are found in nature is such that the frequency of quantitative occurrences is inversely proportional to the quantity.

Simple inverse proportionality implies that variable Y (Frequency) decreases as X (Quantity) increases, without any powers or exponents involvement, and hence $Y = k/x$, for some constant number denoted by k. The postulate implies that the histogram is falling to the right at a very particular rate and that the height is always being reduced by half whenever quantity X is doubled.

The constant denoted with the lowercase letter "**k**" in the numerator is interchangeably denoted by the capital letter "**K**" as well.

Frequency $= k/$Quantity

An elegant feature associated with the postulate is that the total quantity is the same for all sizes.

The trade-off between the size (X) and its frequency (k/x) perfectly cancels out under the postulate, so that: Total Quantity $=$ (Size) $\times$ (Frequency) $= (X) \times (k/x) = K =$ Constant.

This fixed-total-quantity feature is not valid for distributions such as k/x^2, k/x^3, or k/x^4 of very steep fall, nor with distributions such as $k/x^{1/2}$, $k/x^{1/3}$, or $k/x^{1/4}$ of very mild fall. Uniquely, this feature is true only when the rate of the fall in the histogram is of the k/x^1 type.

Hence, an alternative and equivalent way to postulate all this could be as follows:

Alternative Postulate: The generic pattern in how relative quantities are found in nature is such that all sizes and quantities are granted equal total quantity and that no size or quantity is given preferred or privileged status.

$$[\text{QUANTITY}_A] \times [\text{FREQUENCY}_A]$$
$$= [\text{QUANTITY}_B] \times [\text{FREQUENCY}_B]$$

Figure 50 depicts one histogram in the spirit of the postulate. Indeed, the height at each point is simply $16/X$. As we move from 1 to 2, doubling the quantity, the height is reduced by half, from 16 to 8. As we move from 2 to 4, doubling the quantity, the height is reduced by half, from 8 to 4. As we move from 4 to 8, doubling the quantity, the height is reduced by half, from 4 to 2.

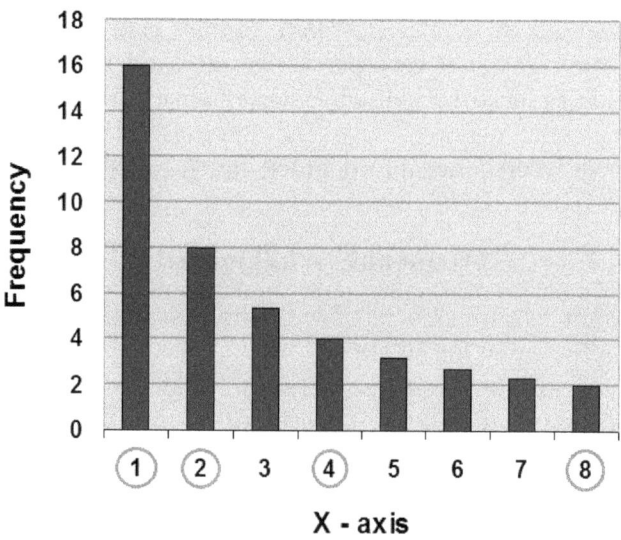

Figure 50. $16/X$: The Doubling of X Reduces Height by Half

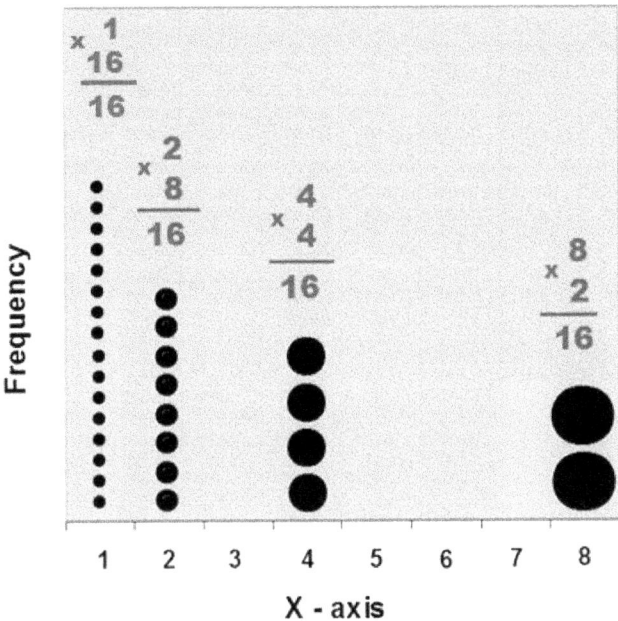

Figure 51. Total Quantity as $(X) \times$ (Frequency) is Constant Throughout $16/X$

The fixed-total-quantity feature can be seen clearly in Figure 51, where the total quantity for size 1 is (Size) × (Frequency) = (1) × (16) = 16; the total quantity for size 2 is (2) × (8) = 16; the total quantity for size 4 is (4) × (4) = 16; and the total quantity for size 8 is (8) × (2) = 16. Surely, bigger sizes might appear as if they contain much more total quantity simply by virtue of being big, but they are also rarer, so there are fewer of them. This trade-off between size and its frequency perfectly cancels out under the postulate. In other words, there exists a perfect balance between size and its frequency, resulting in a constant total quantity for all sizes. Figure 51 and the Alternative Postulate could draw philosophical support from the oval configuration of Figure 15 in chapter 17 which is of the same conceptual approach and where the 3 sizes fairly share equal portions of the overall oval area resulting in decisive quantitative skewness.

Chapter 49

Application of the Postulate via Generic Bin Scheme on k/x

The postulate in the previous chapter shall now be applied deductively in order to arrive at an exact expression for relative bin rank d proportions for all D and F combinations. This is done via explorations of results from (**discrete**) bin schemes of histograms having generic D and F values performed on the abstract (**continuous**) k/x distribution defined from point $P > 0$ to infinity. It is not possible to start the whole scheme from the 0 origin, since $K/0$ is infinite, or rather undefined, and thus $P \neq 0$ is a necessary condition. Moreover, in order to simplify the complex mathematics, a very particular P value is chosen so that $P = w$, namely P is assigned the same value as that of the width of the bins in the first histogram. The separation from the 0 origin to the launch of k/x, denoted by P, should be equal to the width of each bin in the first histogram, which is denoted by w. We start with an infinitesimally small bin width w and let the bins grow gradually via repeated multiplications by F. Hence, the value of P is also exceedingly small, and the launch of k/x is quite near the 0 origin.

The ultimate aim is to evaluate the bin rank d probabilistic areas of k/x for the generic bin scheme constructed with any D and F combinations. The plan is *not* to evaluate and analyze distinct parts of one whole fixed *infinite* scheme already put in place, but instead to construct the k/x curve and the associated bin schemes, cycle by cycle, gradually and in stages, and to evaluate these very limited bin schemes being constructed and evolving on *finite* ranges, growing in scope as the number of cycles increases. Then, we extrapolate further in the abstract the sequence of algebraic expressions coming out of these evaluations as the number of cycles goes to infinity in the limit.

Figure 52 depicts the *first* histogram of the bin scheme constructed for the abstract k/x distribution. In Figure 53, the bin scheme is increased to *two* histograms, and then in Figure 54, the bin scheme is increased to *three* histograms. Each illustration of a finite set of one, two, or three histograms is followed by the calculations of the relative areas of the bins, representing the preliminary probabilities of bin ranks via basic calculus on these limited cycles without going to infinity. The ultimate or proper probabilities are those calculated when an infinite number of such histograms are considered and consolidated, at least in a limiting sense via extrapolations.

1 Histogram on K/X

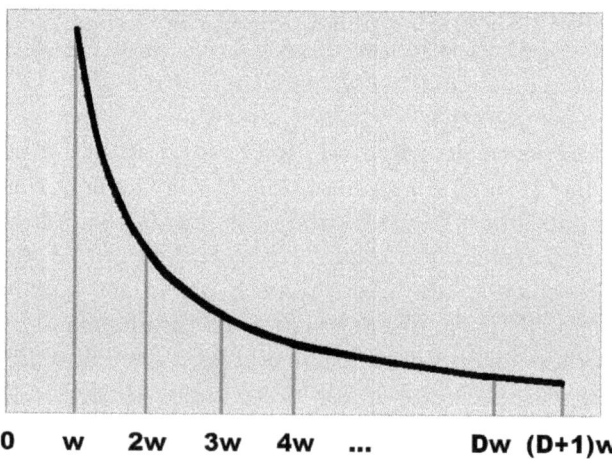

0   w   2w   3w   4w   ...   Dw   (D+1)w

Figure 52.   The First Histogram of an Infinite Bin Scheme for k/x

Equating the entire area in Figure 52 to one, i.e., to 100%, as in all statistically proper densities, we obtain

$$\int_{w}^{(D+1)w} \frac{k}{x} \, dx = 1$$

$[k] \times [\ln((D+1)w) - \ln(w)] = 1$
$[k] \times [\ln(D+1) + \ln(w) - \ln(w)] = 1$
$[k] \times [\ln(D+1)] = 1$
$k = 1/\ln(1+D)$

Evaluating the portion of area hanging over bin rank d, we obtain

$$\text{Probability}(d) = \int_{dw}^{(d+1)w} \frac{k}{x} \, dx$$

$P(d) = [k] \times [\ln((d+1)w) - \ln(dw)]$
$P(d) = [k] \times [\ln(d+1) + \ln(w) - \ln(d) - \ln(w)]$
$P(d) = [k] \times [\ln(d+1) - \ln(d)]$
$P(d) = [k] \times [\ln((d+1)/(d))]$

Finally, substituting $1/\ln(1+D)$ for k, we obtain

$$\boldsymbol{P(d) = \ln[(1+d)/d]/\ln(1+D)}$$

In order to fit this algebraic expression into some well-organized format of higher cycles of histograms, which are soon to be calculated, d is conveniently written as $[1 + (d-1)]$, so that Probability(d) can be written as

$$\frac{\ln\left(\frac{[1+(d)]}{[1+(d-1)]}\right)}{\ln(1+D)}$$

Our deliberate choice of equating the value of the separation of the launch of k/x from the 0 origin (denoted by P) to the width of each of the bins in the first histogram (denoted by w) successfully led to the elimination of P and w altogether from the resultant algebraic expression above, leaving it totally independent of P or w, giving the result a flavor of universality and generality and considerately simplifying the mathematics involved.

Figure 53 depicts the bin scheme limitedly constructed for only two histograms. Equating the entire area to one, i.e., to 100%, as in all statistically proper densities, we obtain

2 Histograms on K/X

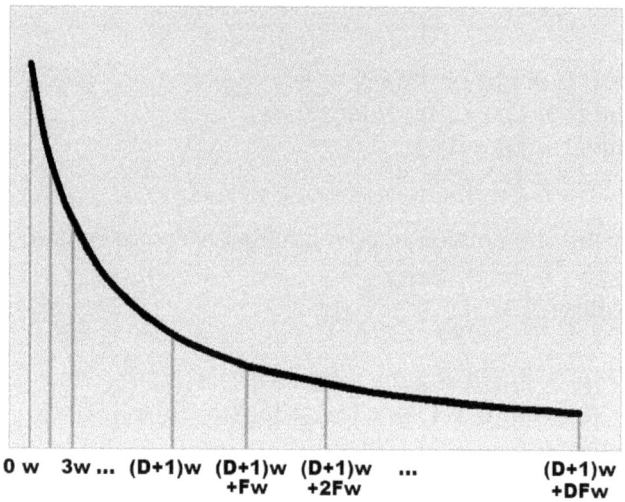

0 w   3w ...   (D+1)w   (D+1)w   (D+1)w   ...   (D+1)w
                        +Fw   +2Fw   +DFw

Figure 53.   The First Two Histograms of an Infinite Bin Scheme for k/x

$$\int_{w}^{(D+1)w+DFw} \frac{k}{x}\, dx = 1$$

$$[k] \times [\ln((D+1)w + (DF)w) - \ln(w)] = 1$$
$$[k] \times [\ln([(D+1) + (DF)]w) - \ln(w)] = 1$$
$$[k] \times [\ln[(D+1) + (DF)] + \ln(w) - \ln(w)] = 1$$
$$[k] \times [\ln[D+1+DF]] = 1$$
$$k = 1/\ln(1 + D + DF)$$

Evaluating the *first* portion of the area hanging over bin rank d belonging to the first cycle, we obtain

$$P1(d) = \int_{dw}^{(d+1)w} \frac{k}{x}\, dx$$

$$P1(d) = [1/\ln(1 + D + DF)] \times [\ln((d+1)w) - \ln(dw)]$$
$$P1(d) = [1/\ln(1 + D + DF)] \times [\ln(d+1) + \ln(w) - \ln(d) - \ln(w)]$$
$$P1(d) = [1/\ln(1 + D + DF)] \times [\ln(d+1) - \ln(d)]$$
$$P1(d) = [1/\ln(1 + D + DF)] \times [\ln((d+1)/d)]$$

Evaluating the *second* portion of the area hanging over bin rank d belonging to the second cycle, we obtain

$$P2(d) = \int_{(D+1)w+(d-1)Fw}^{(D+1)w+(d)Fw} \frac{k}{x}\, dx$$

$P2(d) = [1/\ln(1 + D + DF)] \times [\ln((D+1)w + dFw) - \ln((D+1)w + (d-1)Fw)]$
$P2(d) = [1/\ln(1 + D + DF)] \times [\ln([(D+1) + dF]w) - \ln([(D+1) + (d-1)F]w)]$
$P2(d) = [1/\ln(1+D+DF)] \times [\ln[(D+1)+dF]+\ln(w) - \ln[(D+1)+(d-1)F] - \ln(w)]$
$P2(d) = [1/\ln(1 + D + DF)] \times [\ln[(D+1) + dF] - \ln[(D+1) + (d-1)F]]$
$P2(d) = [1/\ln(1 + D + DF)] \times [\ln[1 + D + dF] - \ln[1 + D + (d-1)F]]$

Combining both cycles, namely $P(d) = P1(d) + P2(d)$, we get

$P(d) = [1/\ln(1+D+DF)] \times [\ln((d+1)/d) + \ln[1+D+dF] - \ln[1+D+(d-1)F]]$
$P(d) = [1/\ln(1+D+DF)] \times [\ln((d+1)/d) + \ln([1+D+dF]/[1+D+(d-1)F])]$
$P(d) = [\ln((d+1)/d) + \ln([1+D+dF]/[1+D+(d-1)F])]/[\ln(1+D+DF)]$

Now, d is conveniently written as $[1 + (d - 1)]$, so that Probability(d) can be written as

$$\frac{\ln\left(\dfrac{[1+(d)]}{[1+(d-1)]}\right) + \ln\left(\dfrac{[1+D+(d)F]}{[1+D+(d-1)F]}\right)}{\ln(1+D+DF)}$$

Figure 54 depicts the bin scheme limitedly constructed for only three histograms, pertaining to the initial three cycles. For the sake of brevity and better visualization, the gray lines separating the bins within each histogram are not shown.

3 Histograms on K/X

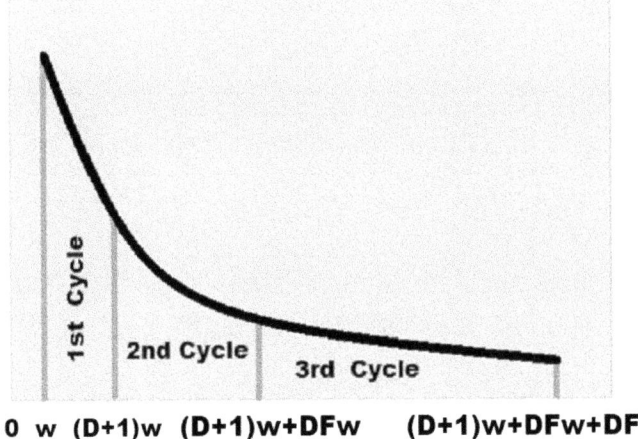

0   w   (D+1)w   (D+1)w+DFw   (D+1)w+DFw+DFFw

Figure 54. The First Three Histograms of an Infinite Bin Scheme for k/x

For the three-cycle scheme, the sum of the relevant three definite integrals is

$$\int_{dw}^{(d+1)w} \frac{k}{x}\,dx + \int_{(D+1)w+(d-1)Fw}^{(D+1)w+(d)Fw} \frac{k}{x}\,dx + \int_{(D+1)w+DFw+(d-1)FFw}^{(D+1)w+DFw+(d)FFw} \frac{k}{x}\,dx$$

Applications of the same type of calculations above as for the cases of one histogram and two histograms lead to the following well-organized final expression:

$$\frac{\ln\left(\frac{[1+(d)]}{[1+(d-1)]}\right) + \ln\left(\frac{[1+D+(d)F]}{[1+D+(d-1)F]}\right) + \ln\left(\frac{[1+D+DF+(d)F^2]}{[1+D+DF+(d-1)F^2]}\right)}{\ln(1+D+DF+DF^2)}$$

Chapter 50

The Infinite Sequence Result for the Bin Scheme on k/x

Straightforward and nearly effortless extrapolations of the expressions of higher cycles enable us to examine their format without laboring long and hard again for each and every set of higher histogram arrangements. In order to be able to decipher (via intense mathematical analysis) the eventual limit of this sequence as the number of cycles goes to infinity, we first arrange and demonstrate the sequence clearly and in an orderly manner. The initial four expressions of probabilistic proportions within each bin rank d for 1-cycle, 2-cycle, 3-cycle, and 4-cycle bin schemes are outlined as follows:

$$\frac{\ln\left(\frac{[1+(d)]}{[1+(d-1)]}\right)}{\ln(1+D)}$$

$$\frac{\ln\left(\frac{[1+(d)]}{[1+(d-1)]}\right) + \ln\left(\frac{[1+D+(d)F]}{[1+D+(d-1)F]}\right)}{\ln(1+D+DF)}$$

$$\frac{\ln\left(\frac{[1+(d)]}{[1+(d-1)]}\right) + \ln\left(\frac{[1+D+(d)F]}{[1+D+(d-1)F]}\right) + \ln\left(\frac{[1+D+DF+(d)F^2]}{[1+D+DF+(d-1)F^2]}\right)}{\ln(1+D+DF+DF^2)}$$

$$\frac{\ln\left(\frac{[1+(d)]}{[1+(d-1)]}\right) + \ln\left(\frac{[1+D+(d)F]}{[1+D+(d-1)F]}\right) + \ln\left(\frac{[1+D+DF+(d)F^2]}{[1+D+DF+(d-1)F^2]}\right) + \ln\left(\frac{[1+D+DF+DF^2+(d)F^3]}{[1+D+DF+DF^2+(d-1)F^3]}\right)}{\ln(1+D+DF+DF^2+DF^3)}$$

227

Chapter 51

The General Law of Relative Quantities

It is necessary to evaluate the limit of the sequence as the number of cycles goes to infinity. Enlisting the help of the distinguished mathematician George Andrews led to fruitful results. Andrews is the world's leading expert on integer partitions, well known also for his discovery of Ramanujan's lost notebook and his subsequent extensive work on its contents. His detailed derivation work shall be added as Appendix A. His reduction of the limit of the infinite sequence to a closed-form analytical expression resulted in the following succinct formula for the aggregated proportions from all the bins of rank d:

$$\text{GLORQ} = \frac{\ln\left(\dfrac{D + d(F - 1)}{D + (d - 1)(F - 1))}\right)}{\ln(F)}$$

where D is the number of bins in each histogram; $F > 1$ is the inflation factor, which is assumed to be greater than 1, signifying actual expansions of the histograms; and lower case d is the bin rank. The notation "ln" refers to the natural logarithm with base e.

The above expression of the limit of the infinite sequence regarding bin schemes constructed for the k/x distribution shall be termed "the general law of relative quantities", with GLORQ as its acronym. The adjective "**general**" is added to indicate that this quantitative phenomenon is the driving force behind the digital phenomenon of Benford's Law, the latter being merely a special case and a consequence of GLORQ. If one chooses to ignore Benford's Law altogether and focus solely on the generic quantitative phenomenon, then "**the law of relative quantities**" would be a more concise and appropriate term.

Andrews' mathematical analysis also shows that for a non-expanding bin system with a flat F, namely when $F = 1$, when all the histograms and their bins are of the same width for all the cycles, the result yields bin equality expressed as $1/D$ for all d.

The moment of truth has arrived. Is the whole conceptual edifice leading up to the infinite sequence correct? Does the postulate truly reflect on how relative quantities occur in physical processes and the natural world? Is the complex mathematical analysis correct? Can GLORQ be considered a physical law of nature? Empirical results from the same 15 real-life data sets and abstract distributions in Figure 48 shall be examined and compared to the theoretical expression above. The theoretical GLORQ expression shall be numerically evaluated for the bin scheme in Figure 48 by substituting in the above formula actual values for the corresponding D and F variables, namely $D = 5$ and $F = 2$. Then, for the empirical part, the average of the 15 bin schemes for these 15 data sets shall be calculated in order to obtain a singular set of proportions representing the empirical result in its entirety. The comparisons between the theoretical and the empirical are as follows:

General Law of Relative Quantities: {26.3%, 22.2%, 19.3%, 17.0%, 15.2%}
Empirical results from 15 data sets: {26.2%, 22.1%, 19.2%, 16.9%, 15.5%}

A remarkable agreement between the theoretical and the empirical is observed, and GLORQ is decisively confirmed! Several other bin schemes with different D and F values performed on these 15 data sets and cases were also examined, and their empirical results also matched the theoretical GLORQ proportions, further validating it. In addition, GLORQ calculated theoretical proportion vectors for the variety of D and F combinations at the two leftmost columns in Figure 49, indeed closely match the result for the earthquake data set shown at the right-hand columns in Figure 49, further validating and confirming the whole theoretical edifice laboriously built here.

The mathematical expression of GLORQ indeed implies the "small is beautiful" phenomenon, and this feature can be verified by mathematically rewriting the expression and demonstrating that the proportion for the higher bin rank $(d + 1)$ is always less than the proportion for the lower bin rank (d). In addition, since GLORQ was obtained by constructing bin schemes for the k/x distribution, it follows that lower bin ranks are always of higher proportions in comparison to higher bin ranks due to the fact that the curve of the k/x distribution itself is consistently and monotonically falling to the right. A straightforward algebraic manipulation of the GLORQ expression yields an alternative format, with the variable d appearing only once in the denominator of the numerator and implying that GLORQ is inversely proportional to d. This is written as follows:

$$\text{GLORQ} = \frac{\ln\left(1 + \dfrac{(F-1)}{D + (d-1)(F-1)}\right)}{\ln(F)}$$

Note: The mathematician Wayne Lawton, in a recent 2023 joint article with the author regarding GLORQ, suggested a somewhat more concise format for GLORQ:

$$\text{GLORQ} = \log_F\left(\frac{D + d(F-1)}{D + (d-1)(F-1)}\right)$$

Chapter 52

Benford's Law as a Special Case and Direct Consequence of GLORQ

It shall now be demonstrated how the mathematical expression for GLORQ directly implies the well-known mathematical expression of $\text{LOG}_{10}(1 + 1/d)$ of Benford's Law for the first digits. But in order to accomplish that, it is necessary to recall how the first-digit distribution of Benford's Law could be interpreted as a particular nine-bin scheme (for the nine first digits) with an inflation factor of 10 (for the base 10), as demonstrated more visually at the end of Chapter 45.

This perspective practically ignores the digits altogether! It focuses on the numerical data itself and its spread over the x-axis. Well, except for the fact that these mini histograms are deliberately constructed over a very particular partition of the entire x-axis range according to the cyclical way first digits occur.

In general, for any positional number system base B with $\{1, 2, 3, \ldots, (B - 1)\}$ as the set of all possible first digits, the bin cycles expand by the inflation factor of B, and the number of bins in each cycle is $(B - 1)$. For example, for our base 10 number system, $F = 10$ and $D = 9$. For the base 6 number system, $F = 6$ and $D = 5$.

In conclusion, first-digit proportions for any positional number system base B are equivalent to the proportions for bin schemes with $\boldsymbol{F} = (B)$ and $\boldsymbol{D} = (B - 1)$. The latter equality could also be written as $B = D + 1$. Hence, for number systems and their first digits, the corresponding bin scheme is one where D is always 1 less than F, or, equivalently, where F is always greater than D by 1, and consequently all such bin schemes tailor-made for positional number systems are constrained by $\boldsymbol{F = D + 1}$.

With such succinct correspondence between these constrained bin schemes of the GLORQ and first-digit distributions for positional number systems, it is now straightforward to demonstrate that digital Benford's Law is simply a special case and a consequence of the GLORQ when bin schemes are constructed under the constraint $F = D + 1$. The term F is then substituted by $D + 1$, and the term $D + 1$ can also be substituted by $BASE$, yielding

$$\mathrm{GLORQ} = \frac{\ln\left(\frac{D+d(F-1)}{D+(d-1)(F-1)}\right)}{\ln(F)}$$

$$= \frac{\ln\left(\frac{D+d(D+1-1)}{D+(d-1)(D+1-1)}\right)}{\ln(D+1)} = \frac{\ln\left(\frac{D+d(D)}{D+(d-1)(D)}\right)}{\ln(D+1)}$$

$$= \frac{\ln\left(\frac{D(1+d)}{D(1+(d-1))}\right)}{\ln(D+1)} = \frac{\ln\left(\frac{1+d}{1+(d-1)}\right)}{\ln(D+1)} = \frac{\ln\left(\frac{1+d}{d}\right)}{\ln(D+1)}$$

$$= \frac{\ln\left(\frac{1}{d}+\frac{d}{d}\right)}{\ln(D+1)} = \frac{\mathrm{LOG}_e\left(1+\frac{1}{d}\right)}{\mathrm{LOG}_e(BASE)} = \frac{\mathrm{LOG}_{\mathrm{BASE}}\left(1+\frac{1}{d}\right)}{\mathrm{LOG}_{\mathrm{BASE}}(BASE)}$$

$$= \frac{\mathrm{LOG}_{\mathrm{BASE}}(1+\frac{1}{d})}{1} = \text{Benford's Law}$$

For our positional number system base 10, the term $BASE$ stands for 10, and we arrive at the expression $\mathrm{LOG}_{10}(1 + 1/d)$. Benford's Law is simply a sideshow to this physical law of nature, which can be measured and detected by ways other than our own digital perceptions.

One highly instructive bin scheme is provided via a nine-bin system with an inflation factor of 10, starting at 0.033 and having an initial bin width of 0.07. This bin scheme only dimly hints at our number system, but definitely not at first digits. Examinations of results from this bin scheme applied to various real-life data sets (where all minimums are larger than 0.033) show a remarkable fit to $\mathrm{LOG}_{\mathrm{TEN}}(1 + 1/d)$. Another nine-bin system with an inflation factor of 10, starting at 0.06 and having an initial bin width of 0.027 (where all minimums are larger than 0.06) also gives nearly identical results. Most significantly, the bins here are not aligned along the digital marks of our number system at all; and each bin within each cycle contains a variety of first digits mixed in, sharing the space of the same bin and living side by side harmoniously, while the resultant overall bin proportion is almost exactly $LOG_{TEN}(1 + 1/d)$. Here, we truly encounter the renown proportions of "Benford's Law" $(30.1\%, 17.6\%, \ldots, 4.6\%)$ in its purest and most general form, without digits, free and independent of any number system whatsoever. This strengthens the perception that $LOG_{10}(1 + 1/d)$ is all about relative quantities and that its digital application is but a minor event in the much larger quantitative drama.

Chapter 53

The Universal Law of Relative Quantities

In our construction of GLORQ, we assumed a fixed and constant inflation factor F, going from one histogram on the left to the next wider histogram on the right, so that the width of the bins grows from one histogram to the next histogram as in steady exponential growth, namely as w, Fw, FFw, $FFFw$, and so forth. Yet, empirical testing of bin schemes on k/x with practically very long ranges (as if of infinite range), as well as on positively skewed real-life data with high OOM, using varying and totally disparate F values, namely a chaotic set of F values which increases or decreases in a totally arbitrary manner, still yields a consistent pattern with fixed and definite d bin proportions across all these Benfordian data sets and the long k/x distribution — so long as that same irregular and arbitrary vector of F factors is fixed and is being used in all the bin schemes. The utilization of a variety of dissimilar inflation factors F for histogram expansions in order to measure skewness in data shall be termed **"irregular bin scheme"**.

The application of diverse and unrelated F inflation factors from each histogram to the next histogram implies that the entire mathematical edifice of the previous chapters collapses upon itself and is not applicable at all, so that the derived mathematical formulas are all totally invalid and irrelevant.

As one concrete example with a resultant clear pattern across several Benfordian data sets and the long-range k/x distribution, empirical testing via computerized programming for an irregular five-bin scheme, starting at the origin 0, having an initial bin width of 0.007, and applying the arbitrary and finite inflation vector of $Fi = \{2, 3, 4, 2, 5, 3, 6, 3, 5, 7, 4, 2, 3, 2, 7, 8, 9, 7, 3, 6\}$, proved highly successful. Although expansion along the x-axis is normally achieved by way of infinitely applying inflation factor(s) without an end, here in practical terms, merely the width of the bin in the last cycle is made sufficiently large to enclose the entire range of each data set, since $(0.007) \times 2 \times 3 \times 4 \times 2 \times 5 \times 3 \times 6 \times 3 \times 5 \times 7 \times 4 \times 2 \times 3 \times 2 \times 7 \times 8 \times 9 \times 7 \times 3 \times 6 > \text{Max}(j)$ of any of the j data set examined. Hence, this very finite net cast over a "small" portion of the x-axis is actually still quite sufficiently wide and long for our purposes. A remarkably steady result is found across all these Benfordian data sets and the long k/x, with the proportion vector of the bin rank d coming out decisively skewed and very close to $\{35.5\%, 22.7\%, 16.9\%, 13.3\%, 11.5\%\}$.

Additional tests applying other combinations of arbitrary choices for the vector of inflation factors Fi also yielded consistent patterns across all these Benfordian data sets (but not for non-Benford data), thus further confirming the phenomenon. This consistent pattern in all irregular bin schemes is termed "the universal law of relative quantities", or ULORQ as its acronym. This is a purely empirical law tested on numerous Benfordian data sets, lacking any mathematical apparatus. A distinct ULORQ bin proportions vector law has to be stated for each distinct Fi inflation vector and a chosen D number of bins.

Chapter 54

Benford Second-Order Digits Interpreted as an Irregular Bin Scheme

On the face of it, GLORQ has its limitation in that it only demonstrates the origin of the first digits phenomenon in Benford's Law, but what about the distributions of the second-order digits, or those of the third and higher orders? Does GLORQ fall short in explaining them? No, it doesn't! And our entire quantitative and number-system-invariant foundation is well equipped to handle and explain higher-order digit distributions in Benford's Law.

The understanding of bin schemes and their ability to measure skewness in statistical densities and data sets facilitates a radically different view of higher-order digit distributions in Benford's Law, allowing us to interpret them as some very particular and irregular arrangements of bin schemes. For example, the second-order digit distribution in base 10 can be viewed as an arrangement of repeated sets of equally spaced and equally-as-wide histograms, each with 10 bins, and without expansion whatsoever between these histograms until an integral power of 10 is encountered, so that the histograms are usually flat, just repeating themselves without any expansion, and $F = 1$, but then occasionally and only rarely, they do actually expand by the factor of $F = 10$, and this occurs only when the first-digit order completes full cycles on IPOT points, such as 1, 10, 100, and 1000. Figure 55 depicts this interpretation.

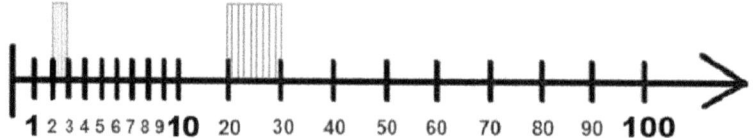

Figure 55.   Second Digits Viewed as Irregular Bins with $F = 1$ and $F = 10$

For example, within each sub-interval $(1, 2), (2, 3), (3, 4), \ldots,$ $(8, 9), (9, 10), (10, 20), (20, 30), \ldots, (80, 90), (90, 100), (100, 200), (200,$ $300)$, and so forth, we cycle a whole 10-bin histogram structure, corresponding to the 10 possible second digits. Within the subinterval $(2, 3)$, for example, as shown in Figure 55, the second digits cycle fully, as in 2.09, 2.18, 2.27, 2.36, 2.44, 2.59, 2.64, 2.72, 2.85, and 2.98, having a bin width of 0.10. Then, the same whole 10-bin histogram structure with the bin width of 0.10 occurs again on the sub-interval $(3, 4)$, and again on $(4, 5)$, and so forth, until we encounter 10. Within the subinterval $(20, 30)$, for example, as shown in Figure 55, the second digits cycle fully, as in 20.34, 21.85, 22.81, 23.29, 24.53, 25.76, 26.49, 27.12, 28.74, and 29.86, having a wider bin width of 1.00. And yet again, on the next subinterval of $(30, 40)$, the second digits cycle fully, as in 30.27, 31.19, 32.63, 33.58, 34.26, 35.41, 36.23, 37.18, 38.27, and 39.87, without experiencing any prior expansion, still having the same bin width of 1.0. It is only beyond 100 that they expand by $F = 10$, widening their bin width to 10.0.

All these allow us to present second digit and higher-order digit proportions as very particular bin schemes too, where, on most occasions, $F = 1$, being flat, but occasionally and less frequently so, they expands by $F = 10$ during the transition of the first-digit cycles on the x-axis. The bin structure of the second digits has a halting/vacillating process of expansion. Surely, the histograms and their bins stay flat for a long while, but then they briefly expand once, then they stay flat for a long time again, and then expand once, and so forth. A great deal of clarity is obtained when a digital second-order distribution is presented simply in terms of a 10-bin scheme with varying inflation factors, Fi. Referring to Figure 55 again, and focusing only on the x-axis part over 1, the vector of Fi in a bin scheme tailor-made for the second-digit order is as follows:

$Fi = \{1, 1, 1, 1, 1, 1, 1, 1, \mathbf{10}, 1, 1, 1, 1, 1, 1, 1, 1, \mathbf{10}, 1, 1, 1, 1, 1, 1, 1, 1,$
$\quad \mathbf{10}, 1, 1, 1, 1, 1, 1, 1, \mathbf{10}, 1, 1, 1, \text{ etc.}\}$

For the digital third order, those $Fi = 10$ occurrences are literally fewer and farther between; $F = 1$ easily outnumbers $F = 10$, and therefore the overall effective F is practically flat. This is why third-digit results show even more digital/bin equality.

One must be reminded that there is no mathematical justification or basis for inserting a singular average F_{AVG} value in the expression of GLORQ for bin schemes with varying Fi values. The whole mathematical edifice laboriously worked out in the previous chapters collapses upon itself when F varies, since it was all based on the constant expansion by F. The simple-minded and naïve extrapolation of the mathematical results of GLORQ with a singular insertion of F_{AVG} value for the bin system interpretation of the second-order distribution in digital Benford's Law with F_{AVG} value of $(1 + 1 + 1 + 1 + 1 + 1 + 1 + 1 + 10)/9 = 18/9 = 2$ and with $D = 10$, applying 2009 Census US population data, leads to:

GLORQ: $D = 10$, $F_{AVG} = 2$:
{13.8, 12.6, 11.5, 10.7, 10.0, 9.3, 8.7, 8.2, 7.8, 7.4}
Benford's Law for second digits:
{12.0, 11.4, 10.9, 10.4, 10.0, 9.7, 9.3, 9.0, 8.8, 8.5}

The two results are a bit similar but not identical. The above result using the average value of Fi actually refers to a 10-bin scheme steadily and consistently expanding by 2, doubling its bin width on each cycle, which is a distinct process from that of the second digits. To truly match the digital second-order distribution in Benford's Law with the bin theory, it is necessary to totally skip any reference to the model of the infinitely expanding bin scheme with fixed F value on k/x, abandoning GLORQ, and turning to ULORQ for help instead. We then simply perform an empirical irregular bin scheme on Benford data sets with the parameters $D = 10$ and alternating F values between $F = 1$ (eight times) and $F = 10$ (once), as above. The result of one such bin scheme performed for the 2009 Census US population data set, pertaining to its 19,509 cities and towns, with a start at the 0 origin, and with an initial bin width of 0.30, is as follows:

ULORQ: D = 10, F = 1, F = 10:
{12.3, 11.4, 10.5, 10.1, 10.2, 9.5, 9.7, 9.0, 9.2, 8.2}
Benford's Law of the second digits:
{12.0, 11.4, 10.9, 10.4, 10.0, 9.7, 9.3, 9.0, 8.8, 8.5}

Indeed, such an irregular bin scheme with 1 and 10 as alternating F values constructed by imitating how the second digits occur on the x-axis yields results that are extremely close to the Benford second-digit proportions. All these demonstrate the consistency and harmony of the conceptual understanding provided by GLORQ and ULORQ, validating the generic approach of the bin schemes.

As expected, the bin theory has just earned one more solid confirmation with this harmonious result. It must be noted that not a single crack of contradiction has ever appeared on any wall in the whole bin edifice whatsoever throughout the entire tour of the palace, not even a small stain on windows or doors with some strange result. Surely, the above irregular bin scheme, which closely fits the law of second digits, has nothing to do with any second-order digits, since a wild mixture of distinct second digits reside harmoniously and cooperatively within each bin and in each histogram! The mild deviations here are only due to the fact that the US population data set is not as ("perfectly") close to Benford as it could be with a finite set of 19,509 data points.

With these latest coherent results, we have developed an understanding of how the confluence of just two features affects the skewness result of any given bin scheme, namely the number of bins D in the system (akin to the base minus 1 in a number system) and either the constant expansion factor F (akin to the base — for the first digits), or the set of varying expansion factors F (akin to the alternating values between 1 and the base — for the second digits). The exact launching point and the exact initial bin width of the first histogram do not affect results almost, so long as the initial bin width is made exceedingly small and as long as the launching point is very near the origin 0 and well below the minimum of all the minimums. Now, we can state conceptually and most broadly the following.

This insight points to the conclusion that base and order in digital Benford's Law are mere variations on bin scheme structure. Namely, that base and order of positional number systems are incorporated into the very fabric of the bin system, constituting its parameters!

Chapter 55

Concluding Historical and Conceptual Perspectives

The crucial generalization of GLORQ over and above Benford's Law is that the $F = D + 1$ restriction of positional number systems is relaxed in order to include the infinitely many possibilities of $F \neq D + 1$ bin schemes as well, and those possibilities cannot be interpreted in any way as positional number systems or first digits. This is why the adjective "general" should perhaps be included in the coining of the term "the general law of relative quantities". This author, who developed the GLORQ theory, still prefers the more concise, simpler, and less pretentious term "the law of relative quantities".

Readers who are interested in a truly comprehensive and methodical study of GLORQ are referred to Section 7 of the author's 2014 book on Benford. The entire section there is imbedded in a science-fiction-like story about a number-less society led by a benevolent dictator who had decreed the abolition of number systems altogether. This was done in order to overcome greed, corruption, and exploitation by simply making it very hard to count or even contemplate large amounts of money via the concept of numbers. The story serves as the backdrop of the whole GLORQ discussion, in part in order to entertain but mostly in order to focus the mind of the reader.

The origin of Benford's Law is all about relative quantities, yet, surprisingly, the story of this phenomenon had been told to us by Simon Newcomb and Frank Benford by way of digits!? Nonetheless, this is not so surprising though, because we do take the *quantitative* pulse within each of the *digital* cycles standing between integral powers of 10, such as 1, 10, 100, and 1000, by way of constructing a nine-bin histogram above each cycle. The digit distribution of the entire data set is then the aggregation of the results coming out of all these nine-bin histograms. For numbers falling exclusively within the subinterval (10, 100), a higher first digit does indeed signify a bigger quantity, as in 85 > 27. And for numbers falling exclusively within the subinterval (100, 1000), a higher first digit does indeed signify a bigger quantity as well, as in 743 > 396.

Our positional number system is surely the most efficient scheme of counting and calculating quantities; and it is typically applied with the decimal base 10, but it encompasses any other bases, such as 2, 8, and 16. Its perceived perfection can be succinctly expressed via our shared belief at this current epoch that there is no hope of discovering another superior system which would ease calculation even further or which would perform better in some new and unsuspected numerical and quantitative ways. We sense that our civilization had arrived at the ultimate number system. A demonstration of the incredible efficiency and perfection of our number system can be obtained and illustrated by pointing out the simple fact that this numerical system hasn't been revised at all in many centuries. It's still the same as it was about 700 years ago; all the while, our modes of cultivation, production, industry, communication, calculation, and transportation have been radically changed and improved. Even while our understanding of mathematics and geometry had deepened considerably, nothing had changed with our positional number system!

In our positional number system, the first digits just happened to cycle in such a way as to perfectly correspond to the particular bin scheme of GLORQ, with $F = D + 1$. One might characterize this correspondence as a coincidence, and certainly the Indian digit-0 inventors Aryabhatta and Brahmagupta in the 5th and 6th centuries, the 8th century Arabic mathematician Al-Khwarizmi who further contributed to the development of our number system, the Italian Fibonacci who popularized the Indo-Arabic numeral system in the Western world, and the others who contributed ideas and paved the way toward a fully functioning positional number system, never attempted or intended to correspond to any bin scheme, nor to discover any consistent statistical pattern in real-life data. Their goal was focused merely on creating an efficient number system to ease counting and calculations. Yet, it so happened that the number system that we invented turned out to have the structure of the bin schemes of GLORQ.

Other number systems, such as Roman numerals or the Egyptian number system, simply do not have the GLORQ bin structure for their first numeral cycles, and this is exactly why when Benfordian data sets are converted into these arcane number systems, no statistical pattern could ever be found regarding their symbols and numerals.

We are no longer seduced and blinded by the incredible efficiency and practicality of our number system, completed during the Renaissance period long ago. We are now able to acknowledge its arbitrariness, and we do not err in believing that the ubiquitous and almost consistent distribution of its symbolic first digits, as in $\mathrm{LOG}_{10}(1 + 1/d)$, constitutes the ultimate in data patterns. The ultimate data pattern is GLORQ!

Appendix A

Infinite Sequence Reduction

This appendix describes the details of the proof produced by the distinguished mathematician George Andrews, leading to a closed-form analytical expression for the limit of the infinite sequence of the bin scheme model for k/x in the $F > 1$ case, as shown in Chapter 50, enabling us to succinctly express the GLORQ.

The fourth term of the sequence expressed earlier, denoted as S_4, is

$$\frac{\ln\left(\frac{[1+(d)]}{[1+(d-1)]}\right) + \ln\left(\frac{[1+D+(d)F]}{[1+D+(d-1)F]}\right) + \ln\left(\frac{[1+D+DF+(d)F^2]}{[1+D+DF+(d-1)F^2]}\right) + \ln\left(\frac{[1+D+DF+DF^2+(d)F^3]}{[1+D+DF+DF^2+(d-1)F^3]}\right)}{\ln(1 + D + DF + DF^2 + DF^3)}$$

Employing the **finite** geometric formula for the terms involving F, namely

$$1 + X + X^2 + X^3 + \cdots + X^N = (X^{N+1} - 1)/(X - 1),$$

the nth term in the sequence is then

$$S_N = \frac{\sum_{j=0}^{N-1} \ln\left(\frac{1 + \frac{D(F^j - 1)}{(F - 1)} + (d)F^j}{1 + \frac{D(F^j - 1)}{(F - 1)} + (d - 1)F^j}\right)}{\ln\left(1 + \frac{D(F^N - 1)}{(F - 1)}\right)}$$

Pulling together all the coefficients of F^{POWER}, we get

$$S_N = \frac{\sum_{j=0}^{N-1} \ln \left(\dfrac{1 + \left(\dfrac{D + (d)(F-1)}{(F-1)} \right) F^j - \dfrac{D}{F-1}}{1 + \left(\dfrac{D + (d-1)(F-1)}{(F-1)} \right) F^j - \dfrac{D}{F-1}} \right)}{\ln \left(1 + \left(\dfrac{D}{(F-1)} \right) F^N - \dfrac{D}{F-1} \right)}$$

In order to obtain a more compact expression, let us define

$$A = \frac{D + d(F-1)}{F-1}$$

$$B = \frac{D + (d-1)(F-1)}{F-1}$$

$$C = \frac{D}{F-1}$$

$$E = 1 - \frac{D}{F-1}$$

$$S_N = \frac{\sum_{j=0}^{N-1} \ln \left(\dfrac{AF^j + E}{BF^j + E} \right)}{\ln(CF^N + E)}$$

Since, in our context, $F \geq 1$, namely either 1, as in flat bin schemes, or larger than 1, as in normal expanding bin schemes yielding quantitative laws, there is no hope of obtaining any obvious convergence in terms such as F^j or F^N, hence we define $f = 1/F$, creating a quantity f such that $0 < f \leq 1$ holds, which may hopefully let terms such as f^j or f^N converge.

$$S_N = \frac{\sum_{j=0}^{N-1} \ln \left(\dfrac{A \dfrac{1}{f^j} + E}{B \dfrac{1}{f^j} + E} \right)}{\ln \left(C \dfrac{1}{f^N} + E \right)}$$

$$S_N = \frac{\sum_{j=0}^{N-1} \ln\left(\dfrac{A\dfrac{1}{f^j} + E}{B\dfrac{1}{f^j} + E} \times \dfrac{f^j}{f^j}\right)}{\ln\left(C \times \dfrac{1}{f^N} + E \times \dfrac{C}{C} \times \dfrac{f^N}{f^N}\right)}$$

$$S_N = \frac{\sum_{j=0}^{N-1} \ln\left(\dfrac{A + E \times f^j}{B + E \times f^j}\right)}{\ln\left(C \times \dfrac{1}{f^N} \times \left(1 + E \times \dfrac{f^N}{C}\right)\right)}$$

$$S_N = \frac{\sum_{j=0}^{N-1} \ln\left(\dfrac{\left(1 + \dfrac{E}{A} \times f^j\right) A}{\left(1 + \dfrac{E}{B} \times f^j\right) B}\right)}{\ln\left(C \times F^N \times \left(1 + \dfrac{E}{C} f^N\right)\right)}$$

$$S_N = \frac{N \times \ln\left(\dfrac{A}{B}\right) + \ln\left(\prod_{j=0}^{N-1} \dfrac{\left(1 + \dfrac{E}{A} \times f^j\right)}{\left(1 + \dfrac{E}{B} \times f^j\right)}\right)}{N \times \ln(F) + \ln(C) + \ln\left(1 + \dfrac{E}{C} f^N\right)}$$

It is only at this late stage that we let N go to infinity!
For $F > 1$ in the normal case of an expanding bin scheme, $0 < f < 1$; therefore,

$$\prod_{j=0}^{\infty} \frac{\left(1 + \dfrac{E}{A} \times f^j\right)}{\left(1 + \dfrac{E}{B} \times f^j\right)}$$

is a convergent infinite product since $\sum_{j=0}^{\infty} f^j$ is converging.

The term $\ln\left(1 + \frac{E}{C} f^N\right)$ is zero as $N \to \infty$. The term $\ln(C)$ is $\ln\left(\frac{D}{F-1}\right)$, and it is also finite as $N \to \infty$. Finally, $N \to \infty$ $S_N = \frac{\ln\left(\frac{A}{B}\right)}{\ln(F)}$, and using the definitions of A and B above, the general relative quantities law is then $\dfrac{\ln\left(\dfrac{\left(\dfrac{D+d(F-1)}{F-1}\right)}{\left(\dfrac{D+(d-1)(F-1)}{F-1}\right)}\right)}{\ln(F)}$, which is

further reduced by canceling out the two $(F-1)$ terms in the numerator to arrive at

The General Law: $\dfrac{\ln\left(\dfrac{D+d(F-1)}{D+(d-1)(F-1)}\right)}{\ln(F)}$

The author made several futile attempts himself to arrive at a closed-form analytical expression for the infinite sequence in the summer of 2013. Enlisting the help of several mathematicians also led to failure. After personally visiting the late mathematician Ralph Raimi (the author of several seminal articles on Benford's Law in the 1960s and 1970s) in Rochester, New York, the author was referred by him to Andrews, who indeed was able to expediently produce the necessary reduction proof in September 2013. Later, Andrews wrote:

> The General Law of Relative Quantities (GLORQ) hinges on a very subtle mathematical limit, and Alex E. Kossovsky enlisted my assistance in its mathematical derivation. I am not an expert on Benford's Law; I am a pure mathematician, however, the subtlety of the limit connected with the GLORQ convinces me that this is a serious subject worthy of much consideration. My experience over the years is that when intricate mathematics is required in a theory, then it often follows that the theory will stand on its own merits. I can assure everybody that the mathematics behind GLORQ is valid and sufficiently surprising, and I believe that any work that has as deep and interesting theorem as its basis merits serious consideration. In any event, I do believe that GLORQ will eventually be accepted.

George Andrews is well known for his discovery of Ramanujan's lost notebook at Trinity College's library, Cambridge University, in 1976 and for his extensive work on Ramanujan's research. In an email to Andrews while lecturing in India in 2023 on Benford's Law and GLORQ, the author asked him whether his reduction achievement was at least partially aided, influenced, or inspired by his decades-long experience working on Ramanujan's mathematics. Andrews replied positively, stating that "[t]he GLORQ proof produced for you was definitely of the Ramanujan world of mathematics." Accordingly, the author indirectly owes gratitude to Ramanujan of India as well.

Appendix B

Data Sets

Exoplanets: Mass of the 1,404 known exoplanets, as of 21 September 2016.

Capitalization: Market capitalization values of 2,889 US companies, as of 9 October 2016.

Population: US 2009 population counts of 19,509 incorporated cities and towns.

Molecules: Molar mass in a list of 2,175 chemical compounds.

Pulsars: Spin frequency of 2,560 known pulsars, as of 2 December 2016.

Expenses: 987,492 expense bills of the State of Oklahoma for the fiscal year 2011.

Price List: The price list of the 14,914 items on sale at Canford Audio PLC.

House Number: 23,633 numbers of home addresses in Prince Edward Island, Canada.

Rock Breaking: Final set with 8,192 pieces of a 100-kilogram rock breaking in 13 stages.

Pipe Breaking: Final set of 13,000 parts from the random real partition of a 38-meter pipe via Uniform.

Exponential Growth: 5% exponential growth series from 7 to 101,643,854 in 338 periods.

Long Chain: 10,000 simulations of the distribution chain of four Uniforms $U(0, U(0, U(0, U(0, 55))))$.

Lognormal: 10,000 simulations from the Lognormal, location 3, shape 1.

Exponential: 10,000 simulations of the Exponential with lambda parameter 666.6.

Earthquakes: Time in seconds between successive earthquakes worldwide for the entire year of 2012, in which 19,452 earthquakes occurred globally.

Appendix C

Glossary of Frequently Used Abbreviations

AIPOT   Adjacent Integral Powers of Ten
CLT   Central Limit Theorem
CPOM   Core Physical Order of Magnitude, $(P_{99\%}/P_{1\%})$
ES04   Excess Sum digits 0 to 4 of the second order
ES12   Excess Sum digits 1 and 2 of the first order
GLORQ   The General Law of Relative Quantities
IPOT   Integral Power of Ten
MCLT   Multiplicative Central Limit Theorem
OOM   Order of Magnitude, $LOG_{10}(\text{Maximum}/\text{Minimum})$
POM   Physical Order of Magnitude, (Maximum/Minimum)
SSD   Sum of Squared Deviations Measure
TES04   Third Excess Sum digits 0 to 4 of the third order

Appendix D

Bibliography

Benford, F. (1938). The law of anomalous numbers. *Proceedings of the American Philosophical Society*, 78, 551.

Benford, F. (2017). Base dependence of Benford random variables. https://arxiv.org/abs/1702.01644.

Benford, F. (2020). Fourier analysis and Benford random variables. https://arxiv.org/abs/2006.07136.

Carslaw, C. (1988). Anomalies in income numbers: Evidence of goal oriented behavior. *The Accounting Review*, 63(2), 321–327.

Cerqueti, R. and Maggi, M. (2021). Data validity and statistical conformity with Benford's law. *Chaos, Solitons and Fractals* (Elsevier Journal), 144, 110740.

Deckert, J., Myagkov, M., and Ordeshook, P. (2011). Benford's law and the detection of election fraud. *Political Analysis*, 19(3), 245–268.

Durtschi, C., Hillison, W., and Pacini, C. (2004). The effective use of Benford's law to assist in detecting fraud in accounting data. *Auditing: A Journal of Forensic Accounting*, 1524–5586/V, 17–34.

Flehinger, J. B. (1963). On the probability that a random integer has initial digit A. *American Mathematical Monthly*, 73(10), 1056–1061 (1966).

Fewster, M. R. (2009). A simple explanation of Benford's law. *The American Statistical Association*, 63(1), 26–32.

Gaines, J. B. and Cho, K. W. (2007). Breaking the (Benford) law: Statistical fraud detection in campaign finance. *The American Statistician*, 61(3), 218–223.

Hamming, R. (1970). On the distribution of numbers. *Bell System Technical Journal*, 49(8), 1609–1625.

Kafri, O. (2009). Entropy principle in direct derivation of Benford's law. http://arxiv.org/abs/0901.3047.

Kossovsky, A. E. (2006). Towards a better understanding of the leading digits phenomenon. http://arxiv.org/abs/math/0612627.

Kossovsky, A. E. (2012). Statistician's new role as a detective — testing data for fraud. http://revistas.ucr.ac.cr/index.php/economicas/article/view/8015.

Kossovsky, A. E. (2013). On the relative quantities occurring within physical data sets. http://arxiv.org/ftp/arxiv/papers/1305/1305.1893.pdf.

Kossovsky, A. E. (2014). *Benford's Law: Theory, the General Law of Relative Quantities, and Forensic Fraud Detection Applications.* World Scientific Publishing Company, Singapore.

Kossovsky, A. E. (2015). Random consolidations and fragmentations cycles lead to Benford's law. https://arxiv.org/abs/1505.05235.

Kossovsky, A. E. (2016). Prime numbers, Dirichlet density, and Benford's law. https://arxiv.org/abs/1603.08501.

Kossovsky, A. E. (2016). Exponential growth series and Benford's law. http://arxiv.org/abs/1606.04425.

Kossovsky, A. E. (2019). Arithmetical tugs of war and Benford's law. http://arxiv.org/abs/1410.2174.

Kossovsky, A. E. (2019). Quantitative partition models and Benford's law. https://arxiv.org/abs/1606.02145.

Kossovsky, A. E. (2021). On the mistaken use of the chi-square test in Benford's law. https://www.mdpi.com/2571-905X/4/2/27.

Kossovsky, A. E. and Lawton, W. (2023). A mathematical analysis of Benford's law and its generalization. https://arxiv.org/pdf/2308.07773.pdf.

Kossovsky, A. E. and Miller, S. (2020). Report on Benford's law analysis of 2020 presidential election data. https://web.williams.edu/Mathematics/sjmiller/public_html/KossoskyMiller_FinalBenfordAnalysis.pdf.

Leemis, L., Schmeiser, B., and Evans, D. (1986). Survival distributions satisfying Benford's law. *The American Statistician*, 54(4), 236–241.

Lemons, S. D. (1986). On the number of things and the distribution of first digits. *American Journal of Physics*, 54(9), 816–817.

Lemons, S. D. (2019). Thermodynamics of Benford's first digit law. *American Journal of Physics*, 87, 787.

Leuenberger, C. and Engel, H.-A. (2003). Benford's law for the exponential random variables. *Statistics and Probability Letters*, 63(4), 361–365.

Miller, S., *et al.* (2008). Chains of distributions, hierarchical bayesian models and Benford's law. http://arxiv.org/abs/0805.4226. *Journal of Algebra, Number Theory: Advances and Applications*, 1(1), 37–60 (2009). https://web.williams.edu/Mathematics/sjmiller/public_html/math/papers/ChainsAndBenford30.pdf.

Miller, S., *et al.* (2018). Benford's law and continuous dependent random variables. https://arxiv.org/abs/1309.5603. *Annals of Physics*, 388, 350–381 (2018). https://web.williams.edu/Mathematics/sjmiller/publ ic_html/math/papers/BenfordDependent85.pdf.

Miller, S., Joseph, I., and Frederick, S. (2013). When life gives you lemons - A statistical model for Benford's law. Williams College. http:// web.williams.edu/Mathematics/sjmiller/public_html/math/talks/small 2013/williams/Iafrate_SummerPoster2013.pdf.

Miller, S., Joseph, I., and Frederick, S. (2015). Equipartitions and a distribution for numbers: A statistical model for Benford's law. Williams College. *Physical Review E*, 91, 6, 062138 (2015), 6 p. http:// journals.aps.org/pre/pdf/10.1103/PhysRevE.91.062138.

Newcomb, S. (1881). Note on the frequency of use of the different digits in natural numbers. *American Journal of Mathematics*, 4(1), 39–40.

Pinkham, R. (1961). On the distribution of first significant digits. *The Annals of Mathematical Statistics*, 32(4), 1223–1230.

Raimi, A. R. (1969). The peculiar distribution of first digit. *Scientific America*, 109–115.

Raimi, A. R. (1976). The first digit problem. *American Mathematical Monthly*, 83(7), 521–538.

Raimi, A. R. (1985). The first digit phenomena again. *Proceedings of the American Philosophical Society*, 129(2), 211–219.

Ross, A. K. (2011). Benford's law, a growth industry. *The American Mathematical Monthly*, 118(7), 571–583.

Sambridge, M., Tkalcic, H., and Arroucau, P. (2011). Benford's law of first digits: From mathematical curiosity to change detector. *Asia Pacific Mathematics Newsletter*. 1(4).

Sambridge, M., Tkalcic, H., and Jackson, A. (2010). Benford's law in the natural sciences. *Geophysical Research Letters*, 37(Issue 22), L22301.

Saville, A. (2006). Using Benford's law to detect data error and fraud: An examination of companies listed on the Johannesburg Stock Exchange. *South African Journal of Economics and Management Sciences* (Gordon Institute of Business Science, University of Pretoria), 9(3), 341–354. http://repository.up.ac.za/handle/2263/3283.

Shao, L. and Ma, B.-Q. (2010). The significant digit law in statistical physics. http://arxiv.org/abs/1005.0660.

Shao, L. and Ma, B.-Q. (2010). Empirical mantissa distributions of pulsars. http://arxiv.org/abs/1005.1702. *Astroparticle Physics*, 33, 255–262.

Stigler, J. G. (1946). The distribution of leading digits in statistical tables. Written in 1945–1946 and referred by Ralph Raimi as a non-published paper.

Varian, H. (1972). Benford's law. *The American Statistician*, 26, 65–66.

Patent: The U.S. Patent Office # 9,058,285. Inventor: Alex Ely Kossovsky.

Titled: "Method and system for Forensic Data Analysis in fraud detection employing a digital pattern more prevalent than Benford's Law".

http://www.google.com/patents/US20140006468.

Date Granted: June 16, 2015.

Index

A

accounting, 8, 34, 64, 67, 140, 158, 255
Achilles' heel of the CLT, 66–67, 96, 178
addition processes, 53–55, 63–68
address data, 104–107
adjacent integral powers of ten, 124, 155, 253
Al-Khwarizmi, 244
alternative postulate, 218, 219
Andrews, George, 229, 247, 250
arithmetical, 64, 68, 75, 96, 256
artificial numbers, 169
Aryabhatta, 244
assigned ID numbers, 169
astronomy, 8
ATM withdrawal numbers, 169
average of averages of averaging scheme, 107, 110
average of averaging scheme, 107

B

Balls and Boxes model, 172
base invariance, 27, 176–178
base invariant, 30, 175, 179
Benford, Frank, 20, 243, 255
Benford, Frank, (grandson), 175, 255
Bin scheme (definition), 212
biology, 9, 61, 116

Boltzmann's entropy formula, 158
Brahmagupta, 244

C

central limit theorem (CLT), 53, 57–59, 253
chains of distributions, 109–116, 151, 170, 256
chaotic repeated partition, 77–79, 84, 87
chemistry, 6, 8, 61, 67, 116, 170
chi-square test, 41–43, 256
classical mechanics, 61
code and index numbers, 169
confidence level, 38, 41
consolidation and fragmentation processes, 95–97
core physical order of magnitude (CPOM), 32–34, 75, 253
curvy-closure, 118, 140–141, 165

D

data aggregation, 20, 103–105, 109, 151, 170
data forensics, 147, 152
data fraud, 147, 182
data fraud detection, 147, 182
data variability, 29, 31
deductive method, 216, 221
deterministic flavor, 145–150, 163, 171

digital development pattern, 146–148, 151–155, 159, 171, 206
dogma, 42, 189
driver's license numbers, 169

E

Earth is not flat, 20
earthquake data, 24, 86–87, 147, 157, 160–161, 177, 214–215, 230
Egyptian numerals, 245
entropy, 159, 173, 256
excess sum digits 0 to 4, 152
excess sum digits 1 and 2, 147, 153
exponential distribution, 81–83, 112, 174, 177
exponential growth, 99–102, 115, 135, 140, 145, 149, 170–171, 173, 237, 251, 256
extrapolation of the second conjecture, 115–116

F

falling log, 130, 138
false negative, 42
false positive, 42
Fibonacci, 244
Fibonacci series, 149, 173
first chain of distribution conjecture, 112, 116
first leading digits, 10–11, 27
first significant digits, 10, 257
first-two-digits, 81, 181–184
Flehinger, Betty, 107–108, 112, 255
forensic digital analysis, 183

G

geology, 8
general form of Benford's law, 125
general law of relative quantities, 205, 229–231, 243, 250, 253, 256
gravitation, 61, 92, 158, 170
gravitational force, 92, 158
Greeks, 189

I

India, 244, 250
inductive method, 214–216
ineffective parameter, 113–115
infinite randomness, 112
infinite sequence, 112, 227, 229–230, 247–250
infinitesimal, 211, 221
inflation factor, 211, 216, 229, 233–234, 237–238, 240
influential parameter, 113–114
integer partitions, 70, 229
integral powers of ten, 124
irregular bin scheme, 237–242

K

k/x distribution, 34–35, 83, 118, 131–135, 145, 149, 155, 157–161, 163, 221–222, 229, 231, 237

L

last-two-digits, 39, 181–184
Lawton, Wayne, 231
law of large numbers, 39
law of relative quantities, 187, 230, 243
Leemis, Lawrence, 143, 156
Lemons, Don S., 89, 172, 256
linear combinations, 64, 121
lognormal distribution, 57–60, 178
lottery numbers, 169

M

Mantissa (definition), 121–127
mean absolute deviations (MAD), 40, 179
Mercury, 210
meta-explanation, 117–118
Miller, Steven J., 89, 116, 256–257
molar mass, 3–6, 10, 13, 64, 67, 251
Mother Nature, 77–78, 84, 118, 140, 157, 159, 173
multiplication processes, 20, 47–50, 58–59, 68, 75, 151, 170, 173
multiplication table, 47–50, 53, 58, 86

multiplicative central limit theorem, 57–59

N

Newcomb, Simon, 205, 243, 257
Newton, Isaac, 61, 92, 157–158, 190
normal distribution, 10, 17, 50, 57–59, 66, 83, 164
number of bins, 204, 211–212, 229, 233, 238, 242
number-system-invariant, 199–200

O

order dependency, 15, 153–154
order of magnitude, 8–9, 20, 29–32, 47, 50, 55, 58, 63, 173, 178, 253
outliers, 29, 31–33

P

Pareto distribution, 114
passport numbers, 169
percentiles, 32, 177–179
physical constants, 170
physical order of magnitude, 29–30, 32, 63, 253
physics, 8, 61–62, 116, 158, 170, 172, 190, 256–257
planet and star formations, 91–93
population data, 9, 43, 87, 101, 183, 206, 241–242
positional number system, 11, 17, 27, 34, 38, 40, 159, 176, 187, 189–190, 192, 195, 200, 233–234, 242–244
positive skewness, 18–20, 35, 47, 50, 59, 69–71, 89, 99, 103, 105, 117, 134, 204, 213, 215
postulate on relative quantities, 217–219
Prime numbers, 172, 256
pseudo-mathematical, 156

R

Raimi, Ralph, 250, 257
Ramanujan, 229, 250

random exponential growth series, 99–102
random flavor, 145, 149, 151, 163, 171
random real partition, 81–84, 86–87, 251
random staged partition, 73–77, 84, 87, 177
randomness within randomness, 111
related log conjecture, 137–144, 165, 174
renaissance, 245
rising log, 129, 138
rock breaking, 74, 177, 251
Roman Empire 192
Roman numerals, 189, 192, 195, 201, 245
Ross, Kenneth, 100

S

scale invariance, 23–25, 170
scale invariant, 30
second chain of distribution conjecture, 115–116
second digits, 13, 15, 19, 39, 41, 134, 141–142, 151–153, 240–242
second-digit development, 141–143, 151–154
simple averaging scheme, 105–107, 110
small is beautiful, 17, 20–21, 213, 230
social security numbers, 169
standard threshold of development, 147
statistical mechanics, 158
sum of squared deviations, 37, 153
super random number, 111

T

thermodynamics, 158, 172, 256
third digits, 19–20, 24, 39, 135, 141–142, 151–152, 183–184
third excess sum digits 0 to 4, 152
third-digit development, 141–143, 151–154
three-dimensional geometry, 158

tug of war, 63–67, 96

two-dimensional random partition,
70–71, 85–86

U

unconditional second digits, 15, 133,
142, 152–153

universal law of relative quantities,
237–238

unconditional third digits, 15, 134,
142, 152–153

uniform distribution, 17, 50, 82, 93,
104, 109–110, 131

uniformity of logarithm, 132

uniformity of Mantissa, 121–122, 125,
131–132, 134, 137, 181

W

Wald distribution, 114

Z

zip code numbers, 169

www.ingramcontent.com/pod-product-compliance
Ingram Content Group UK Ltd.
Pitfield, Milton Keynes, MK11 3LW, UK
UKHW031852100225
454934UK00007B/104